PHILOSOPHY AND GEOGRAPHY II

THE PRODUCTION OF PUBLIC SPACE

Philosophy and Geography   *A Peer Reviewed Annual*

Sponsored by the Society for Philosophy and Geography

Editors:

Andrew Light, Department of Philosophy, The University of Montana
Jonathan M. Smith, Department of Geography, Texas A&M University

Editorial Assistant: David Roberts, The University of Montana

Associate Editors: Yoko Arisaka, University of San Francisco; Jean-Marc
Besse, Collège International de Philosophie à Paris; Edward Dimendberg,
University of California Press; Thomas Heyd, University of Victoria; Eric
Katz, New Jersey Institute of Technology; Jonathan Maskit, Denison Uni-
versity; Don Mitchell, University of Colorado; Rupert Read, University
of East Anglia; Richard Schein, University of Kentucky; Joanne Sharp,
University of Glasgow

Volume I: Space, Place, and Environmental Ethics

Volume II: The Production of Public Space

Volume III: The Meaning of Place, forthcoming, December 1998.

Volume IV: Aesthetics of Everyday Life, submission deadline:
September 15, 1998.

Volume V: Moral and Political Dimensions of Urbanism, submission
deadline: September 15, 1999.

See page 251 for submission guidelines.

PHILOSOPHY AND GEOGRAPHY II

THE PRODUCTION OF PUBLIC SPACE

Edited by
ANDREW LIGHT and
JONATHAN M. SMITH

ROWMAN & LITTLEFIELD PUBLISHERS, INC.
Lanham • Boulder • New York • Oxford

ROWMAN & LITTLEFIELD PUBLISHERS, INC.

Published in the United States of America
by Rowman & Littlefield Publishers, Inc.
4720 Boston Way, Lanham, Maryland 20706

12 Hid's Copse Road
Cummor Hill, Oxford OX2 9JJ, England

British Library Cataloging in Publication information available

ISSN 1090-3771

ISBN 0-8476-8809-7 (cloth : alk. paper)
ISBN 0-8476-8810-0 (pbk. : alk. paper)

Printed in the United States of America

Contents

Illustrations

Figures

Tables

Acknowledgments

The present volume was supported by The University of Montana Foundation Excellence Fund. We are very grateful to President George M. Dennison of The University of Montana for his support of this annual in particular, and the encouragement of a rigorous research climate at The University of Montana in general. We would also like to specially thank James A. Flightner, Dean of the College of Arts and Sciences at The University of Montana, for his exceptional support of this publication, including his provision of equipment, office space, and administrative leadership. We would like to thank the Departments of Philosophy at The University of Montana and the University of Alberta, the Eco-Research Chair in Environmental Risk Management at the University of Alberta, and the Geography Department at Texas A&M University for additional support throughout the year. At Rowman & Littlefield we thank Christa Acampora (now at the University of Maine), Robin Adler, Deirdre Mullervy, and Jonathan Sisk for their continued work on this annual. A special thanks to David Roberts of the Philosophy Department at The University of Montana who this past year took the position of Editorial Assistant. His efforts have been crucial in the completion of this volume.

As always, we are grateful to our referees who provided comments on the numerous manuscripts that were submitted to this issue. In addition to our Associate Editors, and many members of the Editorial Board, we wish to thank the following scholars for serving as manuscript referees: Laurel Bowman, Robert Burch, Stephen Frenkel, Matthew Gandy, Eric Higgs, Joel Kovel, Steve Moore, Francis Sitwell, and Gary Varner.

Introduction: Geography, Philosophy, and Public Space

Andrew Light and Jonathan M. Smith

Brook Farm was a utopian commune in West Roxbury, Massachusetts, which lasted from 1841 to 1847 and served as the setting for Nathaniel Hawthorne's novel, *The Blithedale Romance* (1852). In the novel, Hawthorne describes the disillusionment of the transcendentalist communist Miles Coverdale. Dispiriting toil in the communal field moved Coverdale to groan that the heavy clods were "never etherialized into thought," but the tillers' "thoughts, on the contrary, were fast becoming cloddish."[1] There is, indeed, a certain tension between thoughts and clods, ideas and earth; between the traditional purview of philosophy and the time-honored commission of geography. It is the tension, Lutwack writes, that leaves humanity "permanently disposed to waiver between acceptance and transcendence of earth, between kinship with earth and revulsion against the environmental dependence that must be suffered equally with animal and vegetable forms of life."[2] *Philosophy and Geography* is founded on a belief that this is a creative tension, that thoughts are not necessarily degraded by the company of clods, that earth and ideas are miscible matters, that geography without philosophy is rather too willing to accept earth as it presently appears, that philosophy without geography is rather too eager to transcend earth and its mire of finitude, limits, and interdependence.

This creative tension is present in this volume on public space, where abstract ideas about the nature of the public mingle with accounts of concrete spaces with which they have been implicated. We begin this introduction with an overview of some of the broad trends and concerns that make the idea and the reality of public space such a timely topic.

The Public Sphere

Many of the articles in this volume make use of some concept of the public sphere, that is, public discussion and criticism of the state, or what

Habermas described as "a domain of our social life in which such a thing as public opinion can be formed."[3] The public sphere was created by the eighteenth century bourgeoisie in the gap that opened between the newly privatized life of individual interests and conscience and the increasingly depersonalized power of the bureaucratic state.[4] Goodman describes it as "the ground that mediates between the private life of individuals as producers and reproducers, and their public roles as subjects and (later) citizens of the state."[5]

A similar idea is sometimes expressed with the more concrete phrase, public square, the place where, according to Bakhtin, history is enacted. It stands in contrast to private spaces such as rooms, houses, and gardens, which are settings for enactment of individual lives. The public square is a site of conflict, a heteroglossia where "the Nietzschean, the peasant or the student speak publicly as such."[6] This conjures a useful image. Picture an open plaza overlooked by a regal balcony. In the plaza stand the people, for the moment listening; on the balcony stands the ruler, for the moment pronouncing. The people assembled in the square become a public once they are able to debate among themselves and respond to the pronouncements of the state with rational protests and formal petitions. In other words, if the balcony is taken as a metonymy of the state speaking to the people, the public square is a metonymy of the people talking among themselves and, perhaps more importantly, talking back to the state.

The purpose of a healthy public square, or sphere, is to foster a form of democracy in which issues emerge from, and are clarified by, this sort of public debate. This is in contrast to the present American system, in which issues are invented and defined by organized interest groups and the role of ordinary citizens is reduced to simply agreeing or disagreeing with these manufactured positions. As a result, formal and informal political institutions seek to make the individual citizen aware of ideals and opinions that he or she is presumed already to hold, and they do little to encourage the idea that politics is a process of creating ideals and opinions that are always, in some sense, in the future.

There is today a widespread feeling that the public sphere is in decline, that rational debate and criticism of the state have deteriorated or disappeared altogether.[7] Explanations vary. The gloomiest might contend that the public has discovered the inefficacy of rational criticism in an age when sensational violence is an effective and perhaps even "legitimate means of communication between the people and their governors."[8] Others blame identity politics and fragmentation of the public into interest groups and lifestyle enclaves.[9] Additional explanations are the retreat of intellectuals into secure university appointments, consolidation of newspaper empires, television's antirational visual culture, and perfection of spin control by corporations and the state. Another explanation, directly relevant to this volume, is the disappearance of the places and "institutions of sociability" where "private individuals gathered to use their reason and form civil society."[10] As Christopher Lasch stated it, "civil life requires settings in which people meet as equals," it requires places that "encourage conversation, the essence of civic life."[11]

Public Space

Public space is in its broadest sense nothing more than space to which all citizens are granted some legal rights of access. This is the sense of the word public expressed in designations such as public lecture, public transportation, public telephone, or, in the United Kingdom, public house. These rights are never absolute. For instance, they are normally limited to the right to occupy public space for a finite time and to engage in certain unavoidable exchanges with the environment. No one questions whether access to a public sidewalk carries with it a right to inhale, exhale, and perspire. More extensive rights are afforded in public spaces intended for visits of longer duration. Sidewalks are not normally furnished with drinking fountains or, in the United States, toilets, whereas most parks are. In remote parks, visitors stay for weeks at a time and are sometimes allowed to gather firewood, food, or material with which to fabricate or furnish lodgings. Much contemporary conflict over public space is in fact conflict over these grades of access directed at individuals who live in spaces, such as sidewalks or parks, that others occupy only briefly. Curiously, the behavior for which they are condemned—curling up on the ground in a sleeping bag, urinating against a tree, cooking over an open fire, neglecting personal hygiene, raising jerrybuilt shanties of tarps and poles—is virtually required in the public space of rural and wilderness parks.[12]

In a second more restrictive sense, public spaces are spaces in which citizens gather to form themselves into, and represent themselves as, a public. Architect Rowan Moore describes them as "spaces where a community acquires a sense of itself."[13] Some writers call these civic spaces, and note that they have disappeared even as spaces open to public access have proliferated.[14] Peter Goheen argues that the streets of nineteenth century Toronto, Canada, were public in this sense. It was not individuals' right of access to the streets that made them public, but the "collective right" of groups to perform in the streets and "influence public opinion through the creation of public events."[15] It is in this sense that, in 1989, Wenceslaus Square, the Brandenberg Gate, and, abortively, Tiananmen Square were public spaces in a way that, say, Yellowstone Park was not.[16]

This is the sense Marshall Berman has in mind when he opines that "the heyday of public space in recent American history was the 1960s."[17]

Berman echoes a widespread sentiment that this sort of public space is in decline, a retrogression that is part of the general diminished ability or willingness of citizens in the West to talk back to their rulers. This decline is not easily separable from conflicts over rights of access or the revulsion of the middle class from spaces they find increasingly threatening and offensive, but it is ultimately more troubling than any amount of vagrancy, litter, or vandalism, because it reveals a growing inability of individuals to imagine themselves as a public.[18]

Perhaps the best way to begin thinking about this is to ask yourself where you would go in the event of a national tragedy such as an assassination. Where would you go to stand together as a public opposed to, say, a military coup d'état? And how would you dress? Art critic Peter Plangens asks, "when was the last time you or a friend dressed in a certain way to be seen in public based on the feeling you were morally obligated to contribute positively to the collective appearance of the public?" Not recently, he suspects, because "the American public has no function when it goes out in public most days."[19]

This is why architect Robert Venturi described the plaza as "un-American." "Americans feel uncomfortable sitting in a square," he said, when "they should be working at the office or home with the family looking at television, or perhaps at the bowling alley."[20] Roberta Smith, art critic for the *New York Times*, notes that "Americans spend very little time relaxing in the company of strangers," and as a people consume public spaces rather like french fries, thoughtlessly and without ceremony or considered pleasure.[21] Perhaps this is because Americans do not, for the most part, enter public spaces seeking knowledge of others or of the larger public. They enter public spaces in pursuit of private ends—not as collective authors of history, but as individual authors of private lives. This is why J. B. Jackson noted that American public spaces are designed to serve "the public as an aggregate of individuals."[22] These spaces grant the individual access to commodities, experiences, or knowledge of nature and history, but they seldom encourage strangers to approach one another.

Many writers interpret this as a culture of denial that feels a morbid, undemocratic repugnance for its own public image.[23] These same writers believe that public spaces should be "educative," a living tableau of "society's inner contradictions," of "economic, racial, and ethnic realities," of "all sorts of people, impulses, ideas, and modes of behavior." In the interest of social realism, they contend, public spaces should be "gritty and disturbing rather than pleasant."[24]

Unsurprisingly, the middle class tends to avoid gritty and disturbing environments given over to an "in your face" culture of social grievance.[25]

This is particularly true for women, many of whom perceive any public space as at least somewhat hostile and constraining.[26] To the great annoyance of social critics, the middle class has for the most part chosen to preserve its social complacency in pleasant places protected by "mechanisms of social filtration."[27] A common and much-studied example is shopping malls, which are attractive, according to Rybczynski, because "they are perceived as public spaces where rules of personal conduct are enforced."[28] An extraordinary and also much-studied example is Disney World, which Venturi described as "nearer to what people really want than anything architects have ever given them."[29]

We are confronted, then, with two ideals: educative public spaces on the one hand and entertaining public spaces on the other, the former grimly realistic and the latter comfortingly fantastic. Educative public spaces are, perhaps, comparable to a metal folding chair, which is undoubtedly a very realistic thing to sit on, insofar as it aids clear thinking about fundamental things like the force of gravity and the distribution of bones and tissue in the human frame. Entertaining public spaces are, alternatively, like a padded recliner with vibrator option. Presented with a choice between the two, the middle-class consumer spends little time in deliberation.

Most Americans today seem to view public space as a source of entertainment, not as a site of civic duty, political ferment, or social education. They consequently evaluate public spaces against other forms of experience manufactured by the entertainment industry, such as movies, television, or the World Wide Web. Entertaining public spaces first appeared in such commercial developments as the Midway (Plaisance) or amusement zone that was appended to the serious exhibitions of the great turn-of-the-century world's fairs. Fully exposed to the whims of consumers and the discipline of cost-benefit analysis in this "hothouse of Midway competition," these artificial environments faithfully reflected the popular taste for sensation, sentiment, and nostalgia.[30] In the twentieth century, retailers and restauranteurs made their establishments increasingly comfortable and entertaining to draw customers away from competitors with otherwise indistinguishable products.[31] The trend was accelerated by movie theaters and amusement parks, commercial spaces that provide industrial-strength entertainment as an enticement to consumers who are asked, essentially, to rent a chair (stationary or moving, as the case may be).

The urban expectations and architectural ideals of the majority of consumers in the developed world have been shaped by these spaces, and by an entertainment industry that has taught them to desire environments that are stimulating, vivid, and safe. As a result, "people *expect* to be entertained by the commercial environment," as well as by most other facets of social life.[32] This has given rise to a galaxy of new semipublic spaces based on shopping and entertainment—the malls, boutiques, festival sites, art centers, cultural districts, heritage zones, entertainment complexes, and sports arenas that are such conspicuous components of the entrepreneurial city.[33]

These expectations have also "put new pressure on other traditional institutions," such as historical monuments and museums.[34] As in the case of the proposed redevelopment of Philadelphia's Independence Mall as an edutainment complex, the imperative for public space is to "draw them in and make the education interesting." To hint that this sort of "Disney America is a bad idea" is to betray an ugly "elitist attitude."[35] Sometimes called "envirotainment," this is the inevitable outcome of competition for consumer's time between physical public space and electronic media, such as television and cyberspace.[36] The transformation is well advanced, although far from completed. The future of public space may be glimpsed, however, in the work of environmental graphic designer Wayne Hunt, who explains that his design for the new New Jersey State Aquarium is meant to reassure visitors that "we're not going to teach you about fish, but we are going to get you excited about fish."[37] The idea that public space should serve as the setting for reasoned debate and rational politics appears likely to slip entirely out of view.

Overview of the Volume

We begin with a symposium on the French Marxist philosopher Henri Lefebvre, surely one of the most important figures at the crossroads of philosophy and geography. In particular, any consideration of the philosophical implications of public space—as a physical and cognitive category—would be incomplete without an analysis of Lefebvre's monumental work of 1974, *The Production of Space*.[38] Though often overlooked by the philosophical community, at least in the English-speaking world, Lefebvre's theory of the three dimensions of spatial practices has had a significant impact on a large number of geographers.[39]

The papers in the symposium were originally presented at a December 1995 meeting of the Society for Philosophy and Geography, held in New York City in conjunction with the Eastern Division Conference of the American Philosophical Association. In the tradition of the Society, papers were presented by both philosophers and geographers, namely, Edward Dimendberg (University of California Press), Neil Smith (Geography, Rutgers University), and Ed Casey (Philosophy, State University of New York at Stony Brook). Fortunately, the audience was also mixed and included distinguished scholars from both fields (e.g., David

Harvey, Arthur Danto, and David Michael Levin), producing an extremely stimulating conversation and informative testament to the continued importance of Lefebvre's work today.

In the first paper of the symposium, Edward Dimendberg argues for the importance of Henri Lefebvre's concept of abstract space in differentiating between a communitarian spatial realm conducive to a robust social life and pseudopublic spaces of consumption and distraction that merely feign social inclusion. The paper begins with an analysis of Lefebvre's debt to Hegel, specifically in the logic of dialectics. Lefebvre's analysis draws on Hegel's concepts of the concrete universal as well as the state as the highest embodiment of reason. Where Lefebvre parts with Hegel is in the former's Marxian refusal to reduce social relations (space) to a product of thought or the concept. Dimendberg goes on to relate Lefebvre's critique of several paradigms of spatial theory: mentalism, textualism, containerism, and activism. Lefebvre's strategy, he claims, is dialectical, incorporating insights of past philosophical positions as moments in the production of knowledge. There are three interrelating dialectics at work in Lefebvre's account: that of concepts, that of spatial terms, and that of spatial history. The second—spatial terms—operates with three of Lefebvre's explanatory terms: spatial practices, representations of space, and spaces of representation. The third explains the historical transitions between different forms of spatiality. Dimendberg concludes by arguing that abstract space is the final term of Lefebvre's dialectic; abstract space is then analyzed in light of the Hegelian abstract.

Neil Smith's "Antinomies of Space and Nature in Henri Lefebvre's *The Production of Space*" examines several central arguments in this important work, focusing on a series of critical antinomies that display both some strengths and some weaknesses in Lefebvre's thinking. The first section discusses Lefebvre's critique of idealism or "mental" conceptions of space, which proceeds by adopting a Marxian emphasis on production and means of production. In this way Lefebvre hoped to reunite physical space, mental space, and social space. In the second section, Smith discusses why Lefebvre settled on space as a pivotal concept. One reason is ontological. Here, Lefebvre draws heavily on Hegel, particularly the notion of the "end of history" and "absolute space" as the "product and residue of historical time." However, Lefebvre offers a parallel historical explanation of the pre-eminence of space, focusing on several decisive shifts that have occurred in the latter half of the twentieth century. Smith professes ambivalence with regard to the union of ontological and historical arguments, claiming that its weakness sharply contrasts with the substance of Lefebvre's discussion of modern abstract space. In the third section, Smith turns to a critique of Lefebvre's rethinking of nature, attempting here to develop a more sustainable framework for the valuable conception of space that Lefebvre offers. Smith claims that, while Lefebvre's central notion, the production of space, "brilliantly theorizes the relativity and relationality of social space, it does so at the cost of returning nature to a pre-Einsteinian state." As such, space has radical priority over nature. Thus, while the production of space is fully acknowledged, Lefebvre disregards the production of nature. In the fourth section, Smith presents the outline of a more unified theory in which the social production of nature implies the production of space. This theory is designed to confront three central weaknesses in Lefebvre's work, including the relation between history and ontology raised in Dimendberg's paper.

Replying to both papers, Ed Casey takes up questions concerning both the role of the concrete universal in Lefebvre's work—brought out in Dimendberg's paper—and the proper assessment of the social role and production of nature, as discussed in Smith's paper. Arguing for a broader conception of space and nature as determined by ideas as much as social relations, Casey points toward a lacuna in the work of Lefebvre, Dimendberg, and Smith: the importance of place proper in the discussion of space. Casey makes a brief but strong case for viewing the ideational history of space as being intrinsically tied to the notion of place and suggests that the concept of place is much more useful in an analysis of the social phenomena discussed in *The Production of Space*.[40] Taken together, all three papers demonstrate the utility of interdisciplinary dialogue on the important bundle of issues involved in any discussion of space.

The remaining papers in the volume are not as closely related to one particular account of public space, but, as with volume 1 of *Philosophy and Geography: Space, Place, and Environmental Ethics*, they represent a broad array of approaches under the general theme of the topic at hand.[41] Taken together, however, all of the contributions offer an extraordinarily thorough portrayal of the issues involved in a critical discussion of public space. Still, as with volume 1, several of the contributors shore up their theoretical contributions with specific case studies on theoretical concerns driven by particular regional problems.

Opening these papers is a contribution to the issue perhaps at the very heart of discussions of public space: its opposition with the private. In "Formal Politics, Meta-Space, and the Construction of Civil Life," Mary Ann Tétreault argues that the division of life space into the public and the private is not only overly simplified, but also represents an important support for the legitimacy of political and social hierarchies. The sharp division between the spheres is what Freud called a "family romance," an ideological vision of community structure that obscures a more complex reality, in this case the existence of a "meta-space" where the public and private spheres interpenetrate and "civil life" occurs. Tétreault uses the ancient Greeks as a case study, examining several facets of Athenian life, including the political, sexual, and economic, demonstrating how the public/private boundary is blurred. The paper concludes with a call for women, who remain politically constrained by the assumption of a public/private split, to construct free spaces within the meta-space in which they can empower themselves.

Challenges to the exclusivity of public space are certainly important. Equally important, however, is the challenge to the existence of what has traditionally been thought of as public space. The next few essays consider such threats, and consistent with the opening symposium, offer both a description and prescription of the problems of the increasing assault on heterogenous space. In "The Stranger on the Green," Luke Wallin uses Barry Greenbie's distinction between proxemic and dystemic spaces to illustrate the devaluation of public space. Proxemic spaces are expressions of local or tribal culture, whereas dystemic spaces are used for "more impersonal, abstract relationships that enable members of various social groups to deal with each other amicably." The paradigmatic example of dystemic space is the town green. Town greens, Wallin holds, are examples of spaces that have lost much of their natural and cultural complexity, as well as their sociological diversity. Today they are simplified, framed, and policed as apolitical viewscapes. Wallin considers two interpretations of this loss: Frederick Jameson's determinist vision of increasing commodity fetishism and Sara M. Evans and Harry C. Boyte's more hopeful treatment of the growing alternatives to traditional dystemic spaces. Wallin concludes that cyberspace and other contemporary dystemic spaces afford some hope for future reclamation of public space.

Next, Ted Kilian critiques two other dominant trends in the literature on public space: first, the definition of public space as a site of impersonal contact and, second, the exclusive socialization of public space through processes of representation. Like Tétreault, Kilian urges that we adopt an approach to public space that does not reify any particular space as "public" or "private." He argues that public and private spaces do not exist as such; "public" and "private" are power relationships that play out in space, and all spaces contain both elements. Kilian suggests examining not only the spaces and the representation constructed within those spaces, but also the power relationships that exist within those spaces that define them as public or private and the users as part of "the public" or as "undesirables" to be removed from the public. Kilian concludes his paper with a closer examination of these power relationships.

Closing this group of papers with "The 'Disappearance of Public Space': An Ecological Marxist and Lefebvrian Approach," John Gulick argues that contemporary discourse about the degradation of public space contain three differing meanings of their subject: "public property," "democratic semiotic space," and the "public sphere." Drawing from an

ecological Marxist adaptation of the work of Karl Polanyi and Henri Lefebvre, Gulick offers a method to make sense of these three different descriptions of public space, and the different intellectual traditions that
inform them. Using an ecological Marxist approach, Gulick applies his
analysis of public space to the problem of residential gentrification in
emergent "global cities."

The next two papers narrow the foregoing focus on general theoretical
issues involved in the idea of public space to a more specific analysis of
regional problems, specifically, those currently manifest in the United
Kingdom. However, not just any space in the U.K. is considered, but
particular kinds of spaces endemic to the local idea of public space. First
is the important notion of the rural.

The 1980s and 1990s have witnessed growing demands on, and competing claims to, the use of rural space. In attempting to support their own
claims, advocates of rural space often invoke notions of what the rural
should be—in the U.K., this is the idea of the "rural idyll." In "Contested
Space: the Rural Idyll and Competing Notions of the Good Society in
the U.K.," Matthew Gorton, John White, and Ian Chaston consider the
competing (and often mutually exclusive) notions of individualist and
communitarian conceptions of justice attached to the nebulous concept
of the rural idyll and suggest that the processes of diversification and fragmentation in rural areas has intensified these conflicts over what should
be considered properly rural. The second section of the paper considers
the re-ordering processes of rural economic diversification, counterurbanization, and agricultural restructuring and their promotion of inter-
and intrarural variations. Gorton, White, and Chaston argue that customary sources of power have been diluted and new forms evolved so that the
policy locality is more heterogeneous and less shapeable by traditional
and individual actors. This has led directly to more diverse normative
arguments concerning the nature of what counts as the rural.

In "The Rights of Rights of Way," Hugh Mason discusses another phenomenon currently prominent in (though not exclusive to) the U.K.: the
sale of public lands to private parties and the subsequent restriction of
public access to those lands. Often such restrictions come in the form of
the traditional "freedom to roam" policy being reduced to particular
"rights of way"—paths or routes restricted in use to particular modes of
travel. Echoing issues raised in Wallin's paper on the Public Green, rights
of way are historically an element of English culture and geography, but
they have not always been protected by statutes of law. As a result, there
are numerous modern difficulties in negotiating a balance between private
land rights and public rights of way. Mason notes some significant points
in the history of rights of way and the development of their legal status
as well as relating several areas of difficulty in their modern definition and

use. The right to rights of way is difficult to classify in the customary framework of human rights. It derives neither from existing legal frameworks nor from a wider moral framework; rather, it seems to derive its power simply from long historical precedent. This specifically spatial right is not easily classified as a negative or protective right, a right to freedom from coercion; nor does it fit neatly into the category of positive rights, rights to be provided with certain basic needs. Mason discusses two questions raised by these considerations: is historical precedent sufficient to establish a right? And, if a right changes its function (negative to positive), is it still properly regarded as the same right?

Continuing the general theoretical analysis of the first few papers and containing some elements of the place specificity of the previous two, the next piece—by John Stevenson, "The Mediation of the Public Sphere: Ideological Origins, Practical Possibilities"—combines a thorough analysis of the importance of Habermas's idea of the "public sphere" with a historical investigation of the structure and function of the public sphere as an ideological space in the United States of the 1940s. Like Lefebvre's *The Production of Space*, Habermas's *The Structural Transformation of the Public Sphere* is a groundbreaking work in the literature on public space and a lightning rod for a particular ideological approach to the subject—the discourse ethics championed by Habermas as the heir to the Frankfurt School.[42] Stevenson shows that there is a good prima facie case for attributing a fairly vibrant Habermasian public sphere to the United States between the wars, especially when compared to the situation in America today. Looking a little deeper though, Stevenson finds several ways in which the 1940s public sphere was crucially constricted and compromised. These difficulties are explicitly treated in the second section, in which Stevenson shows that the problems in the public sphere of the 1940s interrelate with questions raised by some commentators on Habermas and by contemporary work in media and cultural studies: questions of the relative unity or multiplicity of "the" public sphere, and, consistent with Wallin's paper, considerations attendant upon the rise of the new media. These questions raise considerable difficulties for Habermas's narrative of decline. Finally, Stevenson uses this dialectic of chronicle and critique to explore briefly the question of what it might mean to constitute anew a viable public sphere in the contemporary world.

Another theoretical investigation accentuated through regional specific examples is offered in Jean Hillier's, "Representation, Identity, and the Communicative Shaping of Place." Here, drawing on examples from Western Australia, Hillier attempts to demonstrate a multidimensional concept of actors in public participation processes and unpacks their different representations of identity of self and place. Hillier draws out the inherent ambiguities of the meanings and values of such representations and suggests the importance of a negotiated outcome to planning decision processes at a local geographical level. Like Stevenson, the author indicates that a Habermasian-based philosophy of communicative action is of value, but is inadequate here to deal with the reality of the power of participants/actors and their representations. The shaping of public space is the outcome of power struggles between actor networks. Hillier also argues that Habermasian communicative action offers an unstable basis for consideration of difference (concrete otherness) and suggests that a true consensus between different participants with different, often conflicting, representations is extremely unlikely.

Any discussion of space, spatiality, and geography certainly begs the question of the role of maps in the shaping of a peculiarly public conception of space. Of late, even a few philosophers have become interested in the representational properties of maps. The final paper in this volume of *Philosophy and Geography* provides an excellent example of the fruits of interdisciplinary work on this important question. In "Maps and Entitlement to Territory," Mario Pascalev takes up the issue in contemporary international ethics that the idea of "traditional occupation" (simply the idea that occupation of a territory by a national group can serve as the basis of a claim to that territory) is widely perceived as a legitimate moral claim by a people to a piece of territory. Specifically, Pascalev is interested in how such claims are supported by ethnic cartography. Pascalev argues that with respect to ethnic cartography, geographic maps are not objective and scientific tools of representation. Maps project "nationally colored visions of the territorial extent of nations; they are tools central to shaping both national space and a sense of nationalism." Pascalev supports this claim with an examination of nineteenth and twentieth century efforts to map Macedonia, particularly the work of Serbian geographer Jovan Cvijic, who advanced national claims and shifted the balance of international opinion in a matter of twelve years. In the course of his paper Pascalev argues against several claims that attempt to establish the objectivity of map-making, concluding that any human geographer's attempt to produce an unbiased, objective representation of territory necessarily fails. Because of this fact, maps, though they may generate political support, do not uphold a moral claim to territory.

Conclusion

This volume makes it clear that public space is a complex matter, and a fruitful topic of analysis, criticism, and interpretation. What is more, it makes it clear that public space is mutable, subject to regular change in its form, use, and definition. It has never been static. In the mid-fourth century B.C. Diogenes of Sinope consciously violated the conventions of the public spaces of Athens with shocking, dog-like behavior that earned for him the title of cynic. At the end of the nineteenth century poor Americans, urban as well as rural, lost the right to graze their livestock on the garbage and grassy verge of public streets and highways, largely because the animals annoyed and offended an increasingly fastidious middle class. Deviance and convention, exclusion and power: these are persistent components in the perpetual permutation of public space.

There is, however, reason to suppose that change in public space is today occurring with unusual speed, potency, and violence. These rapid, deep, and forcible changes are, to be sure, simply part of the general cataclysm that is permanently rearranging the economies, societies, geographies, and philosophies of all the peoples of this earth; but it is, we feel, an especially important part. It is important because public spaces are ordinary and tangible, and are thus very often the media through which we experience the more general change. Nearly every one of us passes through a variety of public spaces every day. Some of them are physical spaces that we bodily occupy, others are semiotic spaces of public display and communication. It was in these spaces, at some point in the 1980s, that most of us had our first tangible experience of economic restructuring as a weird and troubling juxtaposition of hedonic connoisseurship and grim, pathetic vagrancy. It is in these spaces that most of us have come literally face to face with diversity.

This suggests that, for many of us, experiences in public spaces decide the meaning and value of abstractions such as monetary policy, public responsibility, individual rights, national identity, immigration policy, community solidarity. Public spaces are the citizen's testing ground of social theories and political ideals. This is why they are important. If public spaces are inadequate, or altogether absent, we citizens have no way of knowing what our theories and ideals are worth, or even what they mean.

Notes

1. Nathaniel Hawthorne, *The Blithedale Romance and Fanshawe*, vol. 3 of *The Centenary Edition of the Works of Nathaniel Hawthorne*, ed. William Charvat (Columbus: Ohio State University Press, 1964), 66.

2. Leonard Lutwack, *The Role of Place in Literature* (Syracuse, N.Y.: Syracuse University Press, 1984), 5.

3. Quoted in Frederic Wakeman, "The Civil Society and Public Sphere Debate: Western Reflections on Chinese Political Culture," *Modern China* 19 (1993): 108–38

4. John Nerone, "The History of the Public Sphere," *Journal of Communication* 42 (1992): 163–71.

5. Dena Goodman, "Public Sphere and Private Life: Toward a Synthesis of Current Historiographical Approaches to the Old Regime," *History and Theory* 31 (1992): 1–20.

6. Ken Hirschkop, "Heteroglossia and Civil Society: Bakhtin's Public Square and the Politics of Modernity," *Studies in the Literary Imagination* 23 (1990): 65–75. Advocates of religious viewpoints speak of the naked public square to protest the exclusion of religious opinion from public debate. See Richard John Neuhaus, *The Naked Public Square: Religion and Democracy in America* (Detroit, Mich.: W.B. Eerdmans, 1984).

7. Christopher Lasch, *The Revolt of the Elites and the Betrayal of Democracy* (New York: W. W. Norton, 1995), 161–75.

8. Jaques Barzun, *The Culture We Deserve* (Middletown, Conn.: Wesleyan University Press, 1989), 167.

9. Political philosopher Ronald Dworkin interprets this as an attempt to submerge one's self in "a group that is simply a reflection of the self," to surround this group with "barriers that public opinion can no longer penetrate," and to thereby shelter one's personality "from any socializing forces." Ronald William Dworkin, *The Rise of the Imperial Self* (Lanham, Md.: Rowman and Littlefield, 1995), 65, 195.

10. Goodman, "Public Sphere and Private Life," 7.

11. Lasch, *Revolt of the Elites*, 117, 120.

12. Homeless individuals do not incite the same objections when they live in airports, where their appearance and behavior more closely resemble the public for which the space is intended. Kim Hopper, "Symptoms, Survival, and the Redefinition of Public Space: A Feasibility Study of Homeless People at a Metropolitan Airport," *Urban Anthropology and Studies of Cultural Systems and World Economic Development* 20 (1991): 155–75.

13. Rowan Moore, "Open and Shut," *New Statesman & Society* 3 (12 October 90): 27.

14. Lynn Hollen Lees, "Urban Public Space and Imagined Communities in the 1980s and 1990s," *Journal of Urban History* 20 (1994): 443–65; Alex Krieger, "Reinventing Public Space," *Architectural Record* 183 (1995): 76–77.

15. Peter G. Goheen, "Negotiating Access to Public Space in Mid-Nineteenth Century Toronto," *Journal of Historical Geography* 20 (1994): 430–49.

16. Wakeman, "The Civil Society and Public Sphere Debate."

17. Marshall Berman, "Take It to the Streets: Conflict and Community in Public Space," *Dissent* 33 (1986): 476–85.

18. The historian Eric Hobsbawm (born 1917) describes himself as belonging "to the generation when streets and public places were still called after public men and events," for whom "public events are part of the texture of our lives." He contrasts this to the "eerie phenomena" of "men and women at the century's end" who have grown up "lacking any organic relation to the public past of the times they live in." Eric Hobsbawm, *The Age of Extremes: A History of the World, 1914–1991* (New York: Pantheon, 1994), 4, 3.

19. Peter Plagens, "What Happens When American Art goes Public?" *New England Review* 17 (1995): 58–65.

20. Quoted in James Sanders, "Toward a Return of the Public Place: An American Survey," *Architectural Record* 173 (1985): 87.

21. Roberta Smith, "Art, Food, and Public Space: Some Reflections," *New England Review* 17 (1995): 55–57.

22. J. B. Jackson, "The American Public Space," *The Public Interest* 74 (1984): 52–65. Michael Walzer describes the decline of "open minded spaces," where one is prepared to be interrupted, distracted, and diverted by other people, in "Pleasures and Costs of Urbanity," *Dissent* 33 (1986): 470–75.

23. Don Mitchell, "The End of Public Space? People's Park, Definitions of the Public, and Democracy," *Annals of the Association of American Geographers* 85, no. 1 (1995): 108–33.

24. The quotes are drawn from Berman, "Take It to the Streets," 477; Richard Sennett, in "Whatever Became of the Public Square?" *Harper's* 281 (1990): 49–53. See also, Tim Cresswell, *In Place/Out of Place: Geography, Ideology, Transgression* (Minneapolis: University of Minnesota Press, 1996). For a criticism of the view that public space should be a medium of social criticism, see Nathan Glazer, "Subverting the Context: Public Space and Public Design," *The Public Interest* 109 (1992): 3–21.

25. William Grimes, "Have a #%!&$! Day," *New York Times*, Sunday, 17 October 1993, sec. 9, p. 1. See also, Alfred Kazin, "Big Apple Falling," *New York Times*, Sunday, 26 March 1995, sec. 4, p. 15.

26. Thomas M. Nelson and L. J. Loewen, "Factors Affecting Perception of Outdoor Public Environments," *Perceptual and Motor Skills* 76 (1993): 139–46; Carol Brooks Gardner, "Safe Conduct: Women, Crime, and Self in Public Places," *Social Problems* 37 (1990): 311–28.

27. Louise Mozingo, "Public Space in the Balance," *Landscape Architecture* 85 (1995): 42–47. See also, David Sibley, *Geographies of Exclusion: Society and Difference in the West* (London: Routledge, 1995). On the obvious but frequently overlooked need for public spaces that are pleasant, see Jan Gehl, *Life Between Buildings*, trans. Jo Koch (New York: Van Nostrand Reinhold, 1987).

28. Witold Rybczynski, *City Life: Urban Expectations in a New World* (New York: Scribner's, 1995), 210. Rybczynski continues, "they are more like public streets used to be before police indifference and overzealous protectors of individual rights effectively ensured that *any* behavior, no matter how antisocial, is tolerated."

29. Quoted in Margaret J. King, "Disneyland and Walt Disney World: Traditional Values in Futuristic Form," *Journal of Popular Culture* 15 (1981): 116–140.

30. Barbara Rubin Hudson, "Aesthetic Ideology and Urban Design," *Annals of the Association of American Geographers* 69 (1979): 339–361.

31. Mark Gottdiener, *The Theming of America* (Boulder, Colo.: Westview Press, 1997).

32. Gottdiener, *Theming*, 75.

33. David Harvey, *The Condition of Postmodernity: An Enquiry into the Origins of Cultural Change* (Cambridge, Mass. and Oxford: Basil Blackwell, 1989). Susan M. Roberts and Richard H. Schein, "The Entrepreneurial City: Fabricating Urban Development in Syracuse, New York," *Professional Geographer* 45 (1993): 21–33.

34. Bruce Webb, "All In Fun: Entertainment Architecture," *Texas Architect* 40, no. 5 (1990): 26–33. The reason is what Twitchell describes as the "triumph of the vulgar" that has come with "the fall of the [cultural] gatekeeper and the rise of the accountant." James B. Twitchell, *Carnival Culture: The Trashing of Taste in America* (New York: Columbia University Press, 1992).

35. Mubarak S. Dahir, "The Politics of a Public Space," *Preservation* 48 (1996): 30–31.

36. This should come as no surprise, because today "entertainment is the metaphor for all discourse," on and off the television screen. Neil Postman, *Amusing Ourselves to Death: Public Discourse in the Age of Show Business* (New York: Viking, 1985; New York: Penguin, 1986), 92.

37. Eric La Brecque, "Public Space in the Age of Cyberspace," *Print* 50 (1996): 80–87.

38. Henri Lefebvre, *The Production of Space*, trans. Donald Nicholson-Smith (Cambridge, Mass. and Oxford: Basil Blackwell, 1991); originally published as *La Production de l'espace* (Paris: Anthropos, 1974).

39. Lefebvre attracted the notice of several prominent geographers in the 1980s. See, for example, Edward W. Soja, *Postmodern Geographies: The Reassertion of Space in Critical Social Theory* (London: Verso, 1989); Harvey, *Condition of Postmodernity*, 218–219. Widespread appreciation was not possible, however, until the translation was released in 1991. Geographers have been critical as well as appreciative. See, for example, Michael R. Curry, *The Work in the World: Geographical Practice and the Written Word* (Minneapolis: University of Minnesota Press, 1996), chap. 7.

40. For more on this view, see Ed Casey, *The Fate of Place: A Philosophical History* (Berkeley and Los Angeles: University of California Press, forthcoming). Also look forward to vol. 3 of *Philosophy and Geography*, on the topic of "The Meaning of Place," October 1998.

41. See Andrew Light and Jonathan Smith, eds., *Philosophy and Geography I: Space, Place, and Environmental Ethics* (Lanham, Md.: Rowman & Littlefield Publishers, Inc., 1997).

42. Jürgen Habermas, *The Structural Transformation of the Public Sphere: An Inquiry into a Category of Bourgeois Society*, trans. Thomas Burger, with the assistance of Frederick Lawrence (Cambridge, Mass.: MIT Press, 1989); originally published as *Strukturwandel der Öffentlicheit* (Darmstadt and Neuwied: Luchterhand Verlag, 1962).

Henri Lefebvre on Abstract Space[1]

Edward Dimendberg

Introduction

Of the many ideas developed by the French philosopher, geographer, and social theorist Henri Lefebvre in his 1974 book *The Production of Space*, none is more suggestive than the notion of abstract space with which he characterizes the spatiality of Western capitalist societies since the end of the eighteenth century.[2] Lefebvre successively identified the domain of abstract space with the built environment, a set of social behaviors, an ensemble of cognitive relations, technical procedures, and ideologically coded knowledge. Like so much of his work, the notion of abstract space straddles conventional boundaries between philosophy, geography, social theory, architecture, urban planning, history, and political economy.

At a moment when urban public space is an object of increasing interest and attention, the concept of abstract space offers valuable criteria for distinguishing between spatial realms conducive to a vital collective life and pseudo-public spaces of consumption and distraction that merely feign social inclusion and frustrate the liberating human spatial practice that Lefebvre calls "appropriation."[3] By analyzing those features of abstract space that conceal knowledge of the spatial environment and inhibit the possibilities for a more rewarding spatial experience, we may, I would suggest, arrive at an enhanced definition of a genuinely public "differential" space in Lefebvre's terminology.

Despite the increased attention that Lefebvre's work, especially the once-neglected *The Production of Space*, has received from urban geographers and cultural critics over the past decade, the notion of abstract space has received surprisingly little detailed analysis in its own right. Those commentators who have treated it provide quite divergent summary explanations. For Mark Gottdiener, abstract space realizes "the essential spatial contradiction of society" as the confrontation between "the externalization of economic and political practices originating with the capitalist class *and* the state, and social space, or the space of use values produced by the complex interaction of all classes in the pursuit of every-day life."[4] Michael Dear associates abstract space with "the space of accu-mulation in which the processes of production and reproduction are separated, and space assumes an instrumental function."[5] Derek Gregory understands the specificity of Lefebvre's notion to reside in its emphasis on the "decorporealization," "commodification," and "bureaucratization" of space and its concomitant subservience to "property relations" and "sys-tematic surveillance and regulation by the state."[6] Steve Pile claims that "abstract space transports the body outside itself into a visual regime."[7] And in a recent essay published here in this volume, Neil Smith writes that "abstract space begins to emerge when social labour is differentiated into an abstract form and social wealth thereby aspires to the possibility of infinite mobility, global hegemony. The imposition of abstract space, in which social difference is continually flattened and 'crushed' through the state as well as the economy, is for Lefebvre the hallmark of capitalism."[8]

Each of these understandings of abstract space can be justified by refer-ence to different passages in *The Production of Space*, yet it remains diffi-cult to grasp the specificity of the notion and its relation to these various definitions. Is abstract space merely the spatial manifestation of more pri-mary economic relations, or does it constitute autonomous (or intercon-necting) epistemological, social, and cultural domains? In what sense can it be both global and local? It is no simpler to grasp the interrelation of the distinct meanings of Lefebvre's concept, a task that is complicated by his unsystematic and inconsistent presentation of it in that book and other texts as well. The relation of abstract space to Lefebvre's Hegelian-Marxist account of the production of space also requires clarification if one is to solve what Smith calls "an unresolved contradiction between ontology and history [that] drives much of Lefebvre's vision."[9]

How does abstract space relate to Hegel's understandings of the ab-stract and the concrete? If, as Lefebvre suggests, critical knowledge (*con-naissance*) presupposes a view of space as a "concrete universal," how is Hegel's method particularly suited for grasping the fragmentary and dehistoricized qualities of abstract space? In what manner should one in-terpret his appeal to dialectic and claim that contradictions extend to the spatial as well as the temporal realm? Does the notion of the "production of space" refer to the actual spatial relation, to our knowledge of it, or to both? My goal in this essay is to explicate the Hegelian underpinning of Lefebvre's spatial ontology and to suggest its relevance for understanding the notion of abstract space developed in *The Production of Space*.

Four Paradigms and Three Dialectics

There are four paradigms that Lefebvre repeatedly criticizes and rejects in his book as inadequate foundations for a critical spatial theory. One might call the first paradigm mentalism, the belief that space can be adequately understood as a philosophical concept, if not fully reduced to one. Although grasped by means of conceptual schemes and epistemological categories, space is irreducible to them, and he rejects any theory that conflates space with the activity, product, or form of thought.[10] Kantian idealism or its transcendental interpretation of Euclidean geometry might explain certain features of space, yet their emphases on a priori transcendental forms of sensibility inherent in the subject elide the presence of history, including the history of their own production as concepts.

The second paradigm criticized by Lefebvre is textualism, the idea that space can be assimilated to signification or meaning, a position that emerges in the work of Roland Barthes and other urban semiologists.[11] Even legible spaces are imperfectly understood when approached as predetermined meanings, rather than as objects and sites of production. In this rejection of the linguistic analogy as a model of cultural analysis (including his renowned hostility to structuralism) and his dialectical understanding of language as emerging through interaction between subjects, spaces, and practical activity, Lefebvre's materialist view of language and his effort to reintroduce production into the genesis of discourse can be clearly discerned.[12]

Nor should space be understood as an empty container, a passive and inert holder for social relations, the built environment, cultural meanings, and political confrontation. Criticizing conceptions of space as metaphysically pregiven following from the Aristotelian category of *topos*, the *res extensa* of Descartes, Newton's void, and the absolute of Leibniz, Lefebvre develops a critique of what one might call containerism.[13] This entails the belief that space is essentially empty and static, a passive and inert vessel for other forms of political and economic conflict, especially those generated by the circulation of capital and elevated to primacy by economistic Marxism.[14]

The opposite of containerism is what one could term activism, and Lefebvre wages a vehement attack against understandings of space as presence, instrumental reason, and the visible effects of political hegemony, a mode of domination over people and things he expresses through the French word *savoir*.[15] Space exceeds what is present and visible and can exercise its effects through abstract representations and prohibitions in addition to its more tangible physical manifestations. As we shall later discover when considering his anti-ocularcentrism, an understanding of space can never be derived from merely perceiving it.

Rather than simply reject these prior conceptions, Lefebvre is attempting to formulate a spatial theory that syncretically incorporates earlier philosophical positions as moments in the production of a more adequate form of knowledge. This effort to engage the work of earlier thinkers and glean useful insights from them is what I shall call his dialectic of spatial theories.

Lefebvre's account of abstract space involves no less than three dialectical movements, and any effort to understand his theory must explicate the dynamics of their interrelation. One should not expect a mechanical or formal spatial dialectic in *The Production of Space*, for, rather than a rigid schema, it describes a looser framework within which the production of space—and spatial knowledge—emerges. The second movement, which I will label the dialectic of spatial terms, operates with the three principal explanatory terms proposed in *The Production of Space*: spatial practice, representations of space, and spaces of representation.[16] The third, the dialectic of spatial history, explains the historical transition between different forms of spatiality.

The Dialectic of Spatial Terms

Spatial practice (*pratique spatiale*) is *social* practice embodied in the modalities through which human beings live in space and produce and reproduce themselves within it. When Lefebvre claims that "every society—and hence every mode of production with its subvariants—produces a space, its own space," he is not suggesting that every social order fabricates the ultimate substance or entity properly studied by physics (*POS*, 31). As Edward Soja explains, "Space itself may be primordially given, but the organization, use, and meaning of space is a product of social translation, transformation and experience."[17]

Nor does Lefebvre claim that space exists merely as a collective representation in the minds of ancient Greeks, medieval peasants, or urbanites whose spatial experience would be simply conterminous with their beliefs and representations. Lefebvre is not an idealist, and he insists that the resistance and contribution of material nature inevitably determine space, albeit never without considerable social or cultural mediation. At times he even appears to suggest that nature ceases to exist within capitalist society, and its problematic status within his book has received considerable attention from recent commentators.[18]

Spatial practice encompasses the reproduction of the social relations of production, particularly the division of labor, the interaction between people of different age groups and genders, the biological procreation of the family, and the provision of future labor power. It includes the material production of the necessities of everyday life (houses, cities, roads) and the accumulated knowledge with which societies transform their spatial and social environments. In a tantalizingly brief and undeveloped aside to which I shall later return, Lefebvre borrows Noam Chomsky's notions and suggests that the cohesion of spatial practice "implies a guaranteed level of competence and a specific level of performance" "of each member of a given society's relation to that space" (*POS*, 33).

"Representation of space" (*représentation de l'espace*) is the second notion in the dialectic of spatial terms, defined as "conceptualized space, the space of scientists, planners, urbanists, technocratic subdividers and social engineers, as of a certain type of artist with a scientific bent—all of whom identify what is lived and what is perceived with what is conceived . . . This is the dominant space in any society (or mode of production). Conceptions of space tend . . . towards a system of verbal (and therefore intellectually worked out) signs" (*POS*, 38–39).

These are cognitive entities, conceived spaces and concepts, which belong to the history of ideologies and play a role in social and spatial practice involving the deployment of knowledge and power. Characteristic representations of space include Euclidean geometry, social scientific discourses on urban and social behavior, architectural and planning theories of the built environment, the Quattrocentro theory of visual perspective, Einsteinian relativity theory, and philosophical theories of space of Leibniz, Kant, and Hermann Weyl. But the relationship between inside and outside, constitutive of much architectural signification, is also a representation of space, as is the relation of "frontality" that Lefebvre views as crucial to the fascist aesthetic of monumentality and facades.

The third element in the dialectic of spatial terms is the "space of representation" (*espace de représentation*) that Lefebvre defines as "the dominated—and hence passively experienced—space which the imagination seeks to change and appropriate. It overlays physical space, making symbolic use of its objects. Thus spaces of representation may be said, though again with certain exceptions, to tend towards more or less coherent systems of non-verbal symbols and signs" (*POS*, 39). Spaces of representation are "lived" (*vécu*) environments, and this concept underscores the impossibility of experiencing space without cultural mediation. Lefebvre's emphasis on the lived temporality of space suggests a debt to phenomenology and the work of such thinkers as Bergson, Minkowski, Heidegger, Bachelard.[19] Once of a mythical, animistic or spiritual character, and later associated with particular cities (especially Venice), art works, religious imagery, festival, and such distractions of the culture industry as cinema, spaces of representation shape our subjective experience and imagination.

The three elements in Lefebvre's triad of spatial terms do not constitute an abstract model or a mechanical schema for deriving one term from another. Although necessarily interconnected, they constitute a coherent whole only under favorable circumstances, such as Greek civilization or the period in the history of the Western town from the Renaissance to the nineteenth century (*POS*, 40). Lefebvre pointedly emphasizes his goal of apprehending space concretely (rather than immediately) by means of a theory capable of mediating between ideational, social, and political-economic structures. This unitary spatial theory would overcome the current fragmentation that divides the study of space between disciplines such as geography, architecture and urban planning, sociology, and philosophy.

The Dialectic of Spatial History

The third spatial movement in *The Production of Space* is what I will call the dialectic of spatial history. The transition between the absolute, historical, abstract, and differential spatial forms defines Lefebvre's understanding of history. Each of these forms has different spatial and social practices as its content.

Absolute space, the locus of religion and holy or cursed spaces, is built upon sites of nature (caves, mountain tops, rivers) selected for their intrinsic attributes. Once these sites have been consecrated by priestly castes, their qualities enable a society that has "preserved and incorporated bloodlines, family, unmediated relationships—but [has] transposed them to the city, to the political state founded on the town" (*POS*, 48). Filled with beings and symbols, absolute space introduces distinctions between the full and the empty, speech and writing, and the prescribed and the forbidden. It is the space of death, of funerary monuments and mausoleums, but also the space of a sanctified inwardness that Lefebvre finds evident in Greek temples and Shinto sanctuaries. The space of representation predominates in absolute space, yet paradoxically as soon as it is conceptualized its significance wanes and vanishes.

Not merely a collection of sites and signs, absolute space is "at once indistinguishably mental and social, which *comprehends* the entire existence of the group concerned" (*POS*, 240). Lacking distinctions between public and private (and therefore the category of everyday life he understands as a constitutive trait of modernity), Lefebvre's model for absolute space is the alleged internal unity of Greek civilization and harmony between its artistic, religious, and political forms. This recalls Hegel's understanding of ancient Greece as the embodiment of "immediate spirit," for the concept of absolute space, like Hegel's ideal description of social unity, presents a social formation without highly developed forms of social individuality. In this society of predetermined roles "time and space remain inseparable, the meaning of each was to be found in the other, and this *immediately* (i.e. without intellectual mediation)" (*POS*, 241).[20]

Driven by what Lefebvre describes as "an internal dialectic," absolute space becomes relativized: its naturalness is smashed by the forces of history, and "upon its ruins [are] established the space of accumulation" that he calls historical space (*POS*, 49). This tendency appears in the late imperial or the medieval period, a secularized version of the political and social space of Rome. Its locus is the historical town, now in control of the surrounding countryside and its productive resources. Exchange value becomes general through the circulation of gold and silver.[21] Relational networks of markets and communication systems are established in historical space, a process that, according to Lefebvre, promotes increased social differentiation and a growing abstraction of daily life.

Abstract Space

Abstract space, the final term in Lefebvre's historical dialectic and the negation of absolute and historical space, exemplifies Lefebvre's understanding of abstraction as the hallmark of modernity, a consequence of the industrial and political revolutions of the eighteenth century and the increased prevalence of exchange relations.[22] "It was during this time that productive activity (labour) became no longer one with the process of reproduction which perpetuated social life; but in becoming independent of that process, labour fell prey to abstraction, whence abstract social labour—and *abstract space*" (*POS*, 49).

The absorption of use by the medium of exchange, a tendency exemplified in the transformation from a system of artisanal or handicraft manufacturing to production for the market, defines one key feature of abstract space, but scarcely conveys the full range of Lefebvre's understanding of it.[23] Opposed to the "concrete" space of everyday users, their lived subjective experience of spatial practice (*POS*, 362), abstract space corresponds to an economic mode of life in which "reproducibility, repetition, and reproduction of social relationships" (*POS*, 120) attains precedence over nature. A growing permeation of social life by exchange relations (a consequence of the labor expended in commodity production) yields an increasingly mediated mode of spatiality in which abstraction and differentiation alter traditional relations of production and connections to nature.[24]

The inhibition of spatial practice and spaces of representation in abstract space results in a society whose economic and spatial forms lack the immediacy of earlier social formations. Lefebvre suggests that any space appropriated by human beings will inevitably engage both their bodies (spatial practice) and their imaginations (spaces of representation). Spatial immediacy is not the absence of mediated space—the perfectly neutral spatial void Lefebvre holds to be an impossibility—but rather a domain of space not overwhelmed by technology, abstract conceptualizations, and visual stimuli that insures parity of the body and the imagination with cultural representation.

Dominated by representations of space and conceptual models, abstract space exemplifies the famous pronouncement of Lefebvre's associate, Guy Debord, in *The Society of the Spectacle* (1967) that "All that was once directly lived has become mere representation."[25] Although recent scholarship has emphasized Debord's debt to Lefebvre, it seems likely that the latter actually derived the notion of abstract space from Debord's book where it appears (most likely for the first time) in the following context:

> The Capitalist production system has unified space, breaking down the boundaries between one society and the next. This unification is also a process, at once extensive and intensive, of trivialization. Just as the accumulation of commodities mass-produced for the abstract space of the market inevitably shattered all regional and legal barriers, as well as those corporative restrictions that served in the Middle Ages to preserve the quality of craft production, so too was it bound to dissipate the independence and quality of *places*. The power to homogenize is the heavy artillery that has battered down all Chinese walls.[26]

Here we can discern the principal components of Lefebvre's notion, the unification and homogenization of a space dominated by the mass production of commodities and a diminution of the independent identities of places. In abstract space, spatial practice and spaces of representation recede in importance. The primary tendency in abstract space is the disappearance of the qualitative and its replacement by quantitative practices of measurement, exchange, and calculation. These are exemplified in the mass production, economic rationalization, social scientific management of space, urban planning, and bureaucratically controlled mass consumption that increasingly determine the experience of citizens in capitalist societies.[27]

Although associated by Lefebvre with the modern city and the twentieth century built environment of housing projects, highways, shopping centers, monumental plazas, airports, and tourist sites, the scope of abstract space cannot be reduced to any of these, nor exclusively identified with what is perceivable (*POS*, 49–53). Modern architecture, especially the "technicist" work of Le Corbusier and the Bauhaus, is attacked by Lefebvre as an excessively rational and geometrical mode of building whose abstraction and productivism linked industrialization and urbanization in the service of the state (*POS*, 43, 124, 361).

Abstract space contributes to false consciousness and ideology and cloaks differences in illusory coherence of an ideal space opposed to real spatial practice (*POS*, 14, 393). It shares certain traits with the account of reification developed by Georg Lukács in *History and Class Consciousness* (1924), notably the penetration of all aspects of society by commodification, the fragmentation of the object of production, the freezing of time into a delimitable continuum, and the growing "phantom objectivity" of the market that leads people to adopt a "contemplative" impotent stance toward its products.[28] Yet Lefebvre denies that abstract space can be reduced to a "reifying alienation" or defined as an impersonal "pseudo-subject" of modern social space.

Borrowing the term "formant" from acoustic theory, he presents three elements that define abstract space: the geometric formant, the visual formant, and the phallic formant.[29] Through its reduction of natural and social space to the homogeneous Euclidean space of philosophy, the geometric formant guarantees the political and social utility of space.[30] It does this by facilitating the reduction of lived three-dimensional space to two dimensions; the domain of maps, blueprints, graphic representations, highways, trajectories, and itineraries. Geometric homogeneity and its master trope of the grid dominate its representations of space and lead to a confusion between the geometric and the visual in abstract space.

The visual formant involves a series of metaphoric and metonymic substitutions through which writing and the visual surface of the world are mistaken for the truth of space. Explicitly ocularcentric, it neglects spatial experience involving auditory, tactile, and olfactory stimuli. Space, which once "held up to all the members of a society an image and a living reflection of their own bodies" becomes conflated with visualization, the fetishized province of images and spectacle whose privileged locus becomes the female body, fragmented into diverse parts and signs by advertising and the culture industry (*POS*, 111).

Recalling (if not directly appropriating) Heidegger's argument about representation in "The Age of the World Picture" that "Where anything that is has become the object of representing, it first incurs in a certain manner a loss of being," Lefebvre's assertion that "signs have something lethal about them" underscores his belief in the violent character of the introduction of abstraction into nature.[31] He writes: "The violence. . . manifests itself from the moment any action introduces the rational into the real, from the outside, by means of tools which strike, slice and cut . . . For space is also instrumental—indeed it is the most general of tools" (*POS*, 289).

The third formant Lefebvre calls phallic; it enlists vision to realize its symbolism of force, male fertility, and masculine violence. "Phallic erectility bestows a special status on the perpendicular, proclaiming phallocracy as the orientation of space" (*POS*, 287). Abstract space eroticizes leisure spaces, the pleasure districts of the metropolis or the beach, and "isolates the phallus, projecting it into a realm outside the body, then fixes it in space (verticality) and brings it under the surveillance of the eye" (*POS*, 310). A signifier of plenitude and a destructive force capable of filling a spatial lack, the phallic formant is exemplified for Lefebvre in the skyscraper.

Two Versions of Hegelianism and the Spatial Dialectic

Lefebvre's strong intellectual debt to Hegel and the emergence of his thought from the Hegelian and Lukácsean Marxist tradition have been widely acknowledged.[32] Commentators on *The Production of Space* have noted the key role that Hegelian notions of history, dialectic, and totality play in the book. Yet they typically have restricted Hegel's influence upon Lefebvre to the latter's understanding of the emergence of the modern state, as in passages such as the following:

> . . . According to Hegelianism, historical time gives birth to that space which the state occupies and rules over. History does not realize the archetype of the reasonable being in the individual, but rather in a coherent ensemble comprised of partial institutions, groups, and systems (law, morality, family, city, trade, etc.). Time is thus solidified and fixed within the rationality immanent to space. . . What disappears is history, which is transformed from action to memory, from production to contemplation. As for time, dominated by repetition and circularity, overwhelmed by the establishment of an immobile space which is the locus and environment of realized Reason, it loses all meaning (*POS*, 9).

Hegel's conception of the state as the highest embodiment of reason as developed in *The Philosophy of Right* is interpreted both temporally and spatially by Lefebvre.[33] Historical time emerges as an organic culmination of the earlier stages of absolute and historical space. The role of the modern state in the fabrication of abstract space receives equal emphasis:

> . . . The modern state promotes and imposes itself as the stable centre— definitively—of (national) societies and spaces. As both the end and the meaning of history—just as Hegel had forecast—it flattens the social and 'cultural' spheres. It enforces a logic that puts an end to conflicts and contradictions. It neutralizes whatever resists it by castration or crushing (*POS*, 23).

Yet historical time also anticipates a future mode of production that Lefebvre designates as "the production of space" ". . . which is neither state capitalism nor state socialism, but the collective management of space, the social management of nature, and the transcendence of the contradiction between nature and anti-nature . . ." (*POS*, 103). Driven by advances in productive forces, this "end of history" is realized as the overcoming of the "solidified" time and "immobile" space of the modern state and their sublation in a new form of social life.[34]

Parallel to this familiar narrative of Marxist historicism, Lefebvre appropriates another element of Hegel's thought, the doctrine of the concept developed in *The Science of Logic* (1812–1816) and later reworked by Marx in the *Grundrisse* (1857). Although *The Production of Space* continually invokes these notions, most commentators have tended to downplay, if not simply ignore, their function and significance in Lefebvre's text. Among geographers there is a curious refusal to engage Lefebvre's reading of Hegel's *The Science of Logic* (a text of vital significance to Marx, Lenin, Lukács and other thinkers in the Marxist tradition), as if to suggest that Lefebvre's Hegelianism can be explained solely by his reading of *The Phenomenology of Spirit* and *The Philosophy of Right*.[35] I want to argue against this view and to suggest the significance of Hegel's *The Science of Logic* in the spatial ontology that underscores Lefebvre's dialectic of spatial theory, terms, and history, as well as his concept of abstract space.

Lefebvre develops his understanding of the production of space as a "concrete universal" or "concrete generality" (*konkrete Allgemeine*) by appropriating Hegel's discussion of "The Doctrine of the Concept" in *The Science of Logic*.[36] Concrete features of an object are contingent, inessential, and localized in space and time, while abstract traits are essential elements of the identity of a thing that are sharply separated from one another. Logic treats the essence of an entity and exists as structurally prior to its concrete manifestation in a particular instance accessible to sensory perception.

The universal (*das Allgemeine*) for Hegel is the form of absolute self-mediated identity, a repository for the concepts that constitute the I. Thought is the active self-actualizing universal through which the faculty of the understanding abstracts objects from the flow of sense data. Universals are abstract in the sense of being opposed to perception and existing only in the mind as factors of pure non-differentiated identity. They are general concepts that apply to all instances of a given class, such as colored or shaped, and determine the particulars in which they are instantiated. Yet they are also concrete concepts that reconcile opposites and exist as logically prior to particulars and individuals.

Thinking, for Hegel, is the medium in which the universal as the concept (*der Begriff*) differentiates itself into its component moments of the particular (*das Besondere*) and the individual (*das Einzelne*). His doctrine of the concept challenges Kant's view that the I or the understanding abstracts—and remains distinct from—concepts originating in sensory data. As one commentator notes:

> . . . without concepts, there could be no I or understanding . . . Moreover, the I is both wholly universal or indeterminate—if I think of myself simply as a Cartesian ego, bereft of a body or empirical content and particular, in that it cannot exist without a corporeal embodiment and a determinate consciousness of objects other than itself. Thus the structure of the I mirrors that of the concept, which is at once universal, particular, and individual, and which, like the I, embraces ('comprehends') or overreaches (übergreift) what is other than itself.[37]

A particular is a determinate negation of the universal inflected by difference that abstracts or reduces its absolute self-identity or universality, such as a table in the dining room, localized in space and time. The individual is the identity of the universal and the particular. It is the negation of the negation, the particular understood as the unity of identity and difference in becoming, such as the actual dining room table with its full range of attributes. As Inwood suggests, "Hegel rejects the view that universals, particulars and individuals are logically, ontologically, or epistemologically sharply distinct from each other. The universal is concrete, not abstract, and develops into but maintains itself in, the particular and the individual."[38]

A concrete universal contains both the particular and the individual, indeed its concreteness resides in the fact that the three moments of the concept co-exist within it. Understood as a concrete universal, the concept is self or spirit particularized as differences; universal because of its self-identity, concrete (or complete) because it possesses its opposites within itself. For example, rather than specify justice or freedom with abstract definitions Hegel suggests they are concrete universals that develop from *both* the application of an identical law (the universal) and the historical circumstances of their concrete existence, their employment in particular instances.[39]

The relevance of Hegel's dialectical logic to *The Production of Space* appears conspicuously where knowledge is described as the union of the Hegelian concept and subjectivity (*POS*, 6). Substituting "the production of space" for Hegel's "concept," we arrive at Lefebvre's theoretical understanding of socially produced space as a concrete universal containing logical and mathematical generalities, descriptions of particular spaces, and actual individual places accessible to sensory reality (*POS*, 15–16).[40] The concept of space exceeds that which is perceptible (the immediate for Hegel) and harbors a repertoire of forms, representations, and practices that may be only partially evident at any given moment but still constitute its unity as a concrete universal. "Without the *concepts* of space and its production, the framework of power (whether as reality or concept) simply cannot achieve concreteness" (*POS*, 281).

Yet Lefebvre also comprehends existing social space as a concrete universal containing three terms (spatial practice, representations of space, and spaces of representation), three levels (perceived, conceived, lived), and three forms of space (absolute, historical, abstract) that particularize themselves with specific contents at different time periods. Although one probably seeks in vain for an exact correlation between Hegel's terminology and Lefebvre's that would identify specific moments of the concept with different phases of spatial history (universality = absolute space or spatial practice, particularity = historical space or space of representation, individuality = abstract space or representations of space), the fundamental congruence between the Hegelian concept and Lefebvrean space should by now be apparent.

For Lefebvre, space possesses an internal universalizing dynamic, a tendency that it shares with the spatializing (splitting and unifying) movement of Hegel's doctrine of the concept. Although space can be particularized in a representation or as an embodied practice, it remains a concrete unity, any single term of which contains the others, despite the fragmentary appearance of its immediate manifestations. Spatial practice presupposes spaces of representation, such as parks or gardens. The most prosaic representations of space (such as the grid or the curve) manifest geometric and religious concepts of order that serve in part to explain their power. And Lefebvre's discussion of "dialectical centrality" (*POS*, 331–333) suggests that peripheral spaces contained within political or urban centers can challenge established hegemonies.

Lefebvre's dialectic of spatial terms asserts the ontological unity of space (*POS*, 71) and reminds us of the rich theory of spatiality outlined by Hegel in his doctrine of the concept in *The Science of Logic*, a book that Lefebvre presciently and perceptively reads as a contribution to spatial theory. One can only hope that Lefebvre's reading will allow Hegel the recognition he justly deserves as a philosopher of both space and time.

The existence of space as a concrete universal grounds the possibility of our knowledge of it as a concrete totality.[41] Where Lefebvre differs from Hegel is in the refusal he shares with Marx to reduce social relations, here space as a concrete universal, to a product of thought or the concept. The concrete totality of space is the product of thinking, "of the working-up of observation and conception into concepts" to use Marx's phrase, not a self-creating entity or idea.[42] Thought does not actualize the real space whose unity can only be apprehended partially in sensory perception and incorporated in an act of *connaissance* as an element in a totality.

Abstract Space and the Hegelian Abstract

As I have suggested, a key feature of abstract space is its mediated character. Unlike the spatiality characteristic of earlier societies, abstract space presupposes highly developed relations of mass production, commodification, exchange, and instrumental reason. Recalling descriptions of the reified commodity world associated with notions of spectacle, phantasmagoria, and the mass ornament, abstract space is the product of industrial standardization that liquidates earlier spatial, cultural, and architectural forms.[43] An older organic spatiality is replaced by a space in which the social and economic relations of the late capitalist mode of production become visible. Abstract space, like Debord's notion of spectacle, is "capital accumulated to the point where it becomes image."[44]

One paradoxical consequence of this extreme mediation is the reduced concreteness of abstract space. Breaking with Hegel, Lefebvre does not identify spatial mediation with spatial concreteness *tout court*, but suggests how different forms of mediation lead to spatial abstraction. Concrete space resides in the spatially actual and immediate and is closer to the body, although Lefebvre's sketchy remarks on the body in *The Production of Space* beg the question of defining abstract space in relation to an anterior and allegedly "more natural" unity.[45] While he recognizes the multiple cultural, linguistic, and social determinants of the body, Lefebvre assumes (without theorizing) the existence of a primary bodily reality—analogous to the primary spatial reality of differential space—that inevitably reasserts itself against the travails of abstract space.

A second defining trait of abstract space is its existence as simultaneously homogeneous and fragmented, unified and fractured. To borrow the terminology of Hegel, the form of abstract space both particularizes and universalizes its contents by separating and subordinating them to a centralized power. Lefebvre cites as examples a space capable of being pulverized by computer science into discrete bits of data and reassembled as information in separate locations, the dissolving of the village or town into the new modalities of the conurbation, or the geopolitics of air transportation with its web of legal, territorial, and communication practices. Knowledge of this space entails its apprehension as both unified and disjointed:

> At a theoretical level this betrays (rather than simply uncovers) a global strategy; it constitutes a new totality, whose elements appear to be both joined (joined in space by authority and by quantification) and disjoined (disjoined in that same fragmented space and by that same authority, which uses its power to unite by separating and to separate by uniting) . . .[46]

The challenge for Lefebvre's theory is to analyze abstract space as a totality, construing it as neither homogeneous (the error of interpreting it as a transparent reflection of the world as immediacy) nor heterogeneous (the mistake of overlooking how specific contents compose moments of a dialectical unity). If construed as a neutral entity, "external to social practice and hence on those grounds mental or fetishized (objectified)" space cannot be analyzed as a totality (*POS*, 320). "It is not, therefore, as though one had global (or conceived) space to one side and fragmented (or directly experienced) space to the other . . . For space 'is' whole and broken, global and fractured, at one and the same time. Just as it is once conceived, perceived, and directly lived" (*POS*, 356).

As a set of locations where contradictions are generated, the medium that those contradictions tear apart, and the means by which they are hidden by the appearance of consistency, abstract space requires a dialectical explanation that can overcome the tendency of power to "unite by separating and to separate by uniting." How then should we understand abstract space as a concrete universal? Because all social space would by definition contain the universal, particular, and individual moments, abstract space is therefore a defective manifestation of the Hegelian concept with a diminished epistemological and experiential referent. The inability of the bourgeoisie to reduce their spatial practice to abstract space—to assimilate their lived experience to its forms—leads to "spatial contradictions," a deficiency in spatial concreteness (*POS*, 63).

Lefebvre suggests that space is abstract when perceived, conceived, or lived as one-sided and incomplete, the opposite of the Hegelian concrete and the object of an intuition unmediated by concepts and the activity of reason. It can be compared to the abstract universal in Hegel's sense of being a merely determinate empty concept, lacking the full range of individuality and concrete determinations contained within the universal. As Hegel writes in *The Science of Logic*:

> . . . Therefore, what makes this universality abstract is that the mediation is only a condition or is not posited in the universality itself. Because it is not posited, the unity of the abstract universality has the form of immediacy, and the content has the form of indifference to its universality, for the content is not present as the totality which is the universality of absolute negativity. Hence the abstract universal is, indeed, the Notion, yet it is without the Notion; it is the Notion that is not posited as such.[47]

Lefebvre's account of how we might develop knowledge of abstract space exemplifies the Hegelian critique of representational thinking that separates the universal, particular, and individual. Our perception of the world tends to remain trapped on the level of space as immediately conceived and frustrates attempts to draw connections between its different manifestations grasped as a totality. When writing that "abstract space cannot be conceived in the abstract," he recalls Hegel's criticisms of the limitations of the understanding "splitting the concrete into abstract determinatenesses" that remain excluded from the more inclusive syntheses of dialectical and speculative reason. [48] The mediations of politics, economics, culture, and science constitute abstract space, and we can only understand it concretely through them.

Yet the mode of reason that abstract space lacks is not conceptual, but social, and resides in the body and the imagination and those spatial forms that promote their activity. The "abstract universal" of abstract space lacks the moments of corporeal experience and the imaginative activity that comprise concrete space, Lefebvre describes the realm of spatial practice:

> The user's space is lived—not represented (or conceived). When compared with the abstract space of the experts (architects, urbanists, planners), the space of the everyday activities of users is a concrete one, which is to say, subjective. As a space of 'subjects' rather than of calculations, as representational space, it has an origin, and that origin is childhood, with its hardships, its achievements, and its lacks. Lived space bears the stamp of the conflict between an inevitable, if long and difficult maturation process and a failure to mature that leaves particular original resources and reserves untouched. It is in this space that the 'private' realm asserts itself, albeit more or less vigorously, and always in a conflictual way, against the public one (*POS*, 362).

Abstract space is excessively mediated by *representations of space*, but insufficiently permeated by spaces of representation and spatial practice, and indeed could be understood to arrest the development of these moments of the concept of social space. A simple, but powerful example is Lefebvre's remark that trees disappear from abstract space and are replaced by cement structures. Parks or groves of trees permit multiple uses of their space (a picnic, a nap, a lover's tryst, a reading of the daily paper), none of which are formally prescribed or proscribed. Yet a treeless street or shopping mall would frustrate all of these activities and invite only driving or shopping.

Following Lefebvre's appropriation of Chomsky, we might say that abstract space denies spatial users the "performance" of concrete space and substitutes formal rules of spatial "competence" in its place. As a form of spatiality both excessively ideal and conceptual and insufficiently mediated by the standards of social rationality embodied in such earlier forms as the town or the agora, Lefebvre claims abstract space is more prone to violence and war—and more dependent on social consensus—than earlier spatial forms. It realizes the tendency in Western philosophy toward the evacuation of spatial specificity that Edward S. Casey has recently described in his study of the fate of place. [49]

Where's the Dialectic?

Although Lefebvre attributes a weak necessity to the eventual dominance of spatial practice over representation in the production of space, he hedges the question about the necessity of the transition from one moment of spatial history to another, claiming only that historical space was "quickened" by an internal dialectic or that abstract space emerged when labor "fell prey" to abstraction (*POS*, 49).

The absence of historical necessity in the narrative of the transitions between absolute, historical, and abstract space underscores the problematic status of the dialectic in *The Production of Space*. Differential space promotes a spatial politics that proceeds from the body, and an appropriated, rather than dominated, urban environment restores unity to the concrete universal that abstract space breaks up—"the functions, elements, and moments of social practice."[50] The fleeting presence of this utopian concept in *The Production of Space* alludes to Lefebvre's belief in the possibility of a robust form of spatial reason resistant to abstract space and respectful of temporality. Yet apart from a few brief references to the struggles of urban and social groups, Lefebvre has little to say about the agents or spatial practices involved in the emergence of differential space.

Commentators have variously interpreted the transitions between different spatial forms, and their shared suspicion toward the historical foundation of his distinctions between absolute, historical, and abstract space suggests an ultimately ontological intent of his theory.[51] If *The Production of Space* is not a history of space, it is perhaps best described as an ontology of forms and relations of spatiality. I argue that this accurately captures its intellectual agenda and that the ontology and dialectic of spatial terms "ventriloquizes" the dialectic of spatial history.

Lefebvre's reliance on ontology becomes evident when one considers his frequent discussion of spatial "layering" in passages such as the following:

> . . . No space ever vanishes utterly, leaving no trace. Even the sites of Troy, Susa or Leptis Magna still enshrine the superimposed spaces of the succession of cities that have occupied them. Were it otherwise, there would be no 'interpenetration', whether of spaces, rhythms or polarities. It is also true that each new addition inherits and reorganizes what has gone before; each period or stratum carries its own preconditions beyond their limits . . . (*POS*, 164).[52]

What could such spatial interpenetration possibly mean? To claim that no physical space ever completely disappears is a truism that incurs the risk of hypostatizing empty space by separating content and form and leads to the "containerist" view that Lefebvre criticizes. We might also imagine physically devastated spaces, such as military targets or areas ravaged by natural disasters, in which no trace of an earlier space remains discernible. There is consequently good reason to conclude the primary referent of spatial interpenetration is not physical space.[53]

Lefebvre claims social space, especially spaces of representation, to be the privileged domain of spatial interpenetration, as in such passages as the following:

> Be that as it may, the *places* of social space are very different from those of natural space in that they are not simply juxtaposed: they may be intercalated, combined, superimposed—they may even sometimes collide. . . The hypercomplexity of social space should by now be apparent, embracing as it does individual entities and peculiarities, relatively fixed points, movements and flows and waves—some interpenetrating, others in conflict, and so on (*POS*, 88).

Understanding the significance of "spatial architectonics" is a more complex project than Lefebvre suggests, however, for it is not immediately apparent how different spatial regimes can overlap. His metaphor of flaky "Mille-feuille" pastry layers begs the question, as does his criticism of the homogeneous and isotropic spaces of Euclidean and Cartesian mathematics as the basis for a theory of social space (*POS*, 86). The closest answer he provides comes in the form of a discussion of communication networks:

> . . . No space disappears in the course of growth and development: the worldwide does not abolish the local. This is not a consequence of the law of uneven development, but a law in its own right. The intertwinement of social spaces is also a law. Considered in isolation, such spaces are mere abstractions. As concrete abstractions, however, they attain 'real' existence by virtue of networks and pathways, by virtue of bunches and clusters of relationships. Instances of this are the worldwide networks of communication, exchange and information. It is important to note that such newly developed networks do not eradicate from their social context earlier ones, superimposed upon one another over the years, which constitute the various markets: local, regional, national, and international markets; the market in commodities, the money or capital market, the labour market, and the market in works, symbols and signs; and lastly—the most recently created—the market in spaces themselves. Each market, over the centuries, has been consolidated and has attained concrete form by means of a network . . . (POS, 86).

In yet another deployment of Hegelian holism, Lefebvre claims the analysis of social space can only attain concreteness when it is approached as a nodal point embedded in a larger totality of social relationships. To understand the social space of a city or an underdeveloped region in isolation from other clusters—or antecedent forms—is to grasp it as abstract and reified and to misconstrue the object of spatial knowledge.

Herein resides the significance of Lefebvre's dialectic of spatial history. The historical space of the renaissance city sublates the absolute space of the church, just as the abstract space of the late capitalist metropolis preserves an older form of centrality of the piazza. "Each analytic stage deals with a residue left over from the previous stage, for an irreducible element—the substrate or foundation of the object's 'presence'—always subsists" (*POS*, 150). Spatial *Aufhebung*, like its temporal variant, negates, preserves, and transcends.

Yet unlike other French Hegelian philosophers such as Alexandre Kojéve, Lefebvre attaches particular importance to the continuity and overlap between dialectical stages.[54] For without the "irreducible element" that preserves the "ontological substratum" of space, spatial forms could not retain their antecedent modalities. Abstract space, prone as Lefebvre claims to erase previous histories, would remain unchecked by earlier modes of social space and unfettered in the expansion of its reifying and dystopian effects. Although Lefebvre claims that only class struggle can avert the growth of abstract space, his theory endows spatial practices, spaces of representation, and the body with an innate predisposition toward resistance—a self-correcting spatial-political mechanism—that asserts itself in the presence of excessive mediation.[55]

Like Hegel's concrete universal, this spatial substratum contains all moments of the production of space within it and insures that no single moment of the concept of social space predominates. Even the Hegelian "negativity" of abstract space that corrodes traces of earlier spatial forms (*POS*, 370) is incapable of rending the fundamental concrete unity of space, grounded for Lefebvre in the body and spatial practice. This is the sense in which I believe his book outlines an ontology of space, rather than a history of it, yet the very opposition between these terms is poorly suited for grasping the dynamism of spatiality.

We are perhaps now in a position to understand the meaning of the dialectic of spatial terms and history in *The Production of Space*. The absence of teleology and necessity in Lefebvre's dialectic of spatial forms suggests, as I have already claimed, that the work of transition between absolute, historical, and abstract space is accomplished by conflicts and contradictions between the spatial terms (spatial practice, spaces of representation, and representations of space) constituting each form. As a concrete universal, space can never be understood simply by an essential definition or a single concept. Lefebvre writes:

Knowledge falls into a trap when it makes representations of space the basis for the study of 'life', for in doing so it reduces lived experience. The *object* of knowledge is, precisely, the fragmented and uncertain connection between elaborated representations of space on the one hand and representational spaces [spaces of representation,] on the other; and this 'object' implies (and explains) a *subject*—that subject in whom lived, perceived and conceived (known) come together within a spatial practice (*POS*, 230).

The limitations of each spatial form are overcome not through the expression of an immanent "spirit" in Hegel's sense but through the haphazard and contingent interaction between spatial terms that constitutes historical and spatial practice. Although it would be incorrect to discern the activity of reason in this process, we might note the congruence between Lefebvre's theoretical effort to grasp the relation between spatial terms and forms and Hegel's concept of speculative reason that engages the total movement of the concept and reconciles opposites in a higher unity.

The Critique of Abstract Space

Lefebvre's influence upon contemporary geography and cultural analysis has been pervasive and his work continues to appear in English translation and provides a key point of orientation for what has become known as the "spatial turn in cultural studies."[56] His understanding of abstract space as a force that disturbs established spatial forms inspires much of the best contemporary work in urban geography and seems more relevant than ever in an age of rampant development among the newly industrializing countries, continued suburban dispersal in the advanced industrialized nations, and the emergence of powerful new technologies of space such as Global Positioning Systems.[57]

Future attempts to theorize public space might well attend to Lefebvre's critique of the predominance of the visual and conceptual and his redirection of our attention to the spatial practices and spaces of representation of everyday life, a project adumbrated in his famous 1968 book, *Critique de la vie quotidienne*.[58] His persistent emphasis on the necessity of attending to the body and spatial practice suggest the limitations of any urban plan that does not attend to the lived experience of city dwellers.

Perhaps most valuable is Lefebvre's suggestion that collective fantasies and rituals of festival associated with spaces of representation be recognized as essential components of any robust definition of public space as lived experience. Public space "ought to be an opening outwards" (*POS*, 146) and begins with the individual body that is always already spatially positioned and marked by differences of gender, race, and class, determinants about which it must be admitted Lefebvre remains curiously silent in his book.[59] Complementing Lefebvre's emphasis on the body, Smith's praise of spatial scale and Casey's invocation of place develop strategies for transforming abstract space into a less enveloping and hegemonic form of spatiality.[60]

Yet as suggestive as the concept of abstract space remains, I wish to draw attention to a fundamental difficulty it raises. This concerns its dependence on the sketchy and incomplete theory of representation presented in *The Production of Space*. Lefebvre's criticisms of the geometric, optical, and phallic formants admit no criteria for distinguishing between benign and insidious spatial manifestations of these discourses. By presenting them as key traits of abstract space innately complicit with the mechanisms of power, he aligns his theory with the anti-ocularcentric tendency of much recent French postmodernist thought. [61]

The apparent denial of any cognitive dimension to these formants that could generate *knowledge* of space incurs the risk of nihilism and a disbelief in the possibility of abstract rationality in the public realm. Even if one accepts the detrimental consequences that Lefebvre ascribes to the proliferation of the three formants, what form of democratic public culture could forego representations of space and only invoke the body and spaces of representation? Could a truly public space exist without universalizing abstractions? Gregory suggestively compares Lefebvre's discussion of the ravages of abstract space to Habermas's account of the colonization of the lifeworld by the system. Yet unlike the German thinker, Lefebvre provides few hints about the character of a non-alienating social rationality.[62]

A second unsettling question concerns the historical specificity of abstract space and the three formants. Are we to believe their reifying tendencies emerge in the aftermath of the French Revolution and come to the fore in the emerging culture of visual and scientific modernity around 1910, as Lefebvre suggests (*POS*, 290, 302)?[63] Yet if this is the case, what is specifically spatial and not *representational* about abstract space? Would the three formants function otherwise in differential space? One notes an universalizing impulse to Lefebvre's argument that ascribes essential social and political consequences to representational practices criticized by thinkers as diverse as Debord and Heidegger. The force of the concept of abstract space would then become practically indistinguishable from previous critiques of reification and spectacle—each of which entails separate historical and philosophical explanations. Or should we not view the power and originality of Lefebvre's concept to reside in the suggestion that space and representation are inseparable in the modern period?

I would suggest that Lefebvre's theory of abstract space invokes the universalizing generalizations of the three formants in an attempt to describe the abstract spatiality of the capitalist mode of production *in general*. Paradoxically, for a thinker as concerned with the concrete as Lefebvre, *The Production of Space* omits any analysis of a *specific concrete abstraction* of the abstract space of capitalism, what Horvath and Gibson call a study of ". . . the impure, unique, and specific occurrence of a pure, ideal abstraction in a given time and place. In other words, it is an abstraction formed through the union of two distinct levels of abstraction whereby the specific *individuality* of a mode of production and its superstructure is identified."[64]

Shifting between different historical instances of capitalism, yet failing to analyze a particular one in detail, Lefebvre's account of abstract space attains only limited concreteness. Yet as a model of the capitalist space of modernity in general, *The Production of Space* should not, I think, be faulted for its incomplete treatment of the succession of different spatialities and omission of a theory of representation. To chastise Lefebvre for his failure to provide these is to misconstrue the synchronic character of his project. If as Gregory suggests, the intent of *The Production of Space* is to "provide a history of the present," it should come as no surprise that Lefebvre has left the present in his book open. His legacy and example challenge us to produce concrete analyses of the abstract spaces of our own contemporary moment.[65]

Notes

1. An earlier version of this text was presented at a session of the Society for Philosophy and Geography devoted to the work of Henri Lefebvre held at the December 1995 Eastern Division meeting of the American Philosophical Association in New York City. I am grateful to Andrew Light, Neil Smith, Edward Casey, and David Harvey for their stimulating exchange on this occasion and to Edward Casey, Arthur Danto, Martin Jay, Thomas Levin, Andrew Light, Dana Polan, Tom Rockmore, and Hayden White for their incisive comments on a previous draft of this essay. Completion of this article was made possible by a resident fellowship at the University of California Humanities Research Institute. Its superb staff and location in Irvine provided an ideal laboratory for the study of abstract space.

2. Henri Lefebvre, *The Production of Space*, trans. Donald Nicholson-Smith (Cambridge, Mass. and Oxford: Basil Blackwell, 1991). All references to the English edition will be noted in parentheses by the abbreviation *POS* (for *Production of Space*) and followed by the page number.

3. Lefebvre suggests that "appropriation itself implies time (or times) rhythm (or rhythms), symbols, and a practice. The more space is functionalized—the more completely it falls under the sway of those 'agents' that have manipulated it

so as to render it unifunctional—the less susceptible it becomes to appropriation. Why? Because in this way it is removed from the sphere of *lived* time, from the time of its 'users', which is a diverse and complex time" (*POS*, 356). A notable attempt to theorize public space informed by the work of political theorist Claude Leforte is found in Rosalyn Deutsche, "Agoraphobia," in *Evictions: Art and Spatial Politics* (Cambridge: MIT Press, 1996), 269–327. A helpful presentation of recent work on the topic is found in the special issues of *Urban Geography* 17, no. 1–2 (February 15–May 15, 1996) devoted to "Public Space and the City." See Don Mitchell, "Introduction: Public Space and the City" in the first issue for a useful summary. Studies that pursue the topic from the perspectives of architecture and urban planning include Zeynep Çelik, Diane Favro, and Richard Ingersoll, eds. *Streets: Critical Perspectives on Public Space* (Berkeley and Los Angeles: University of California Press, 1994); Michael Sorkin, ed., *Variations on a Theme Park: The New American City and the End of Public Space* (New York: Hill and Wang, 1992); Stephen Carr, Mark Francis, Leanne G. Rivlin, and Andrew M. Stone, *Public Space* (Cambridge, England and New York: Cambridge University Press, 1992).

4. M. Gottdiener, *The Social Production of Urban Space* (Austin: University of Texas Press, 1985), 127.

5. Michael Dear, "Les aspects postmodernes de Henri Lefebvre," *Espaces et Sociétés* 76 (1994): 33–34. Translation mine.

6. Derek Gregory, *Geographical Imaginations* (Oxford and Cambridge, Mass: Basil Blackwell, 1993), 382, 401.

7. Steve Pile, *The Body and the City: Psychoanalysis, Space, and Subjectivity* (London and New York: Routledge, 1996), 163.

8. Neil Smith, "Antinomies of Space and Nature in Henri Lefebvre's *The Production of Space*, page 49 of this volume. Hereafter designated "ASN."

9. Ibid., page 2 of this volume.

10. "On the other hand we have epistemological thought, which constructs an abstract space and cogitates about abstract (logico-mathematical) spaces. Most if not all authors ensconce themselves comfortably enough within the terms of mental (and therefore neo-Kantian or neo-Cartesian) space . . ." (*POS*, 24).

11. For a selection of writings by urban semiologists, see M. Gottdiener and Alexandros Ph. Lagopoulos, eds. *The City and the Sign*, (New York: Columbia University Press, 1986).

12. "Semiotics, we are told, is concerned with the instincts of life and death, whereas the symbolic and semantic areas are the province of signs properly speaking. As for space, it is supposedly given along with and in language, and is not formed separately from language. Filled with signs and meanings, an indistinct intersection point of discourses, a container homologous with whatever it contains, space so conceived is comprised merely of functions, articulations and connections—in which respect it closely resembles discourse. Signs are a necessity, of course, but they are sufficient unto themselves, because the system of verbal signs (whence written language derives) already embodies the essential links in the chain, spatial links included. Unfortunately, this proposed compromise, which sacrifices space by handing it on a platter to the philosophy of language, is quite

unworkable. The fact is that signifying processes (a signifying practice) occur in a space which cannot be reduced either to everyday discourse or to a literary language of texts . . ." (*POS*, 136).

13. "According to most historians of Western thought, Descartes had brought to an end the Aristotelian tradition which held that space and time were among those **categories** which facilitated the naming and classing of the evidence of the senses. The status of such categories had hitherto remained unclear, for they could be looked upon either as simple empirical tools for ordering sense data or, alternatively, as generalities in some way superior to the evidence supplied by the body's sensory organs. With the advent of Cartesian logic, however, space had entered the realm of the absolute. As Object opposed to Subject, as **res extensa** opposed to, and present to, **res cogitans**, space came to dominate, by containing them, all senses and bodies . . ." (*POS*, 1). An important, if scarcely acknowledged source for Lefebvre's ideas about absolute space remains the classic work by Alexandre Koyré, *From the Closed World to the Infinite Universe* (Baltimore, Md.: Johns Hopkins University Press, 1957).

14. For a discussion of traditional Marxist views of space, see Edward W. Soja and Costis Hadjimichalis, "Between Geographical Materialism and Spatial Fetishism: Some Observations on the Development of Marxist Spatial Analysis," *Antipode* 2 (1979): 3–11.

15. "The negative connotation that I feel we are justified in attaching to is the suggestion that such knowledge colludes to some degree with power, that it is bound up, whether crudely or more subtly, with political practice—and hence with the multifarious representations and rhetoric of ideology" (*POS*, 367).

16. Throughout this paper I translate Lefebvre's phrase "Les Espaces de représentation" as "spaces of representation" rather than the unfortunate "representational spaces" chosen by Donald Nicholson-Smith in his otherwise elegant English translation of *The Production of Space*.

17. Edward Soja, "The Socio-Spatial Dialectic," *Annals of the Association of American Geographers* 70 (June 1989): 210.

18. "Natural space is irreversibly gone. And although it of course remains as the origin of the social process, nature is now reduced to materials on which society's productive forces operate." Henri Lefebvre, "Space: Social Product and Use Value," trans. J. W. Freiberg, in *Critical Sociology: European Perspectives*, ed. J. W. Freiberg (New York: Irvington Publishers, 1979), 286. For a critique of Lefebvre's understanding of nature, see Neil Smith, "ASN" in this volume and idem, *Uneven Development: Nature: Capital and the Production of Space* (Oxford and Cambridge, Mass.: Basil Blackwell, 1990).

19. An account of Lefebvre's appropriation of the ideas of Bergson, Minkowski, Bachelard, and Heidegger remains to be written.

20. For a sustained account of his views on Greek antiquity, see Henri Lefebvre, *Introduction to Modernity (Twelve Preludes: September 1959–May 1961)*, trans. John Moore (London: Verso Books, 1995), 86–87, 122, 226–28. Hegel develops his understanding of Greek civilization as immediate spirit in *The Phenomenology of Spirit*, trans. A. V. Miller (New York: Oxford University Press, 1977), 289. A discussion of his understanding of the Greeks is found in Georg Lukács,

The Young Hegel, trans. Rodney Livingstone (London: Merlin Press, 1975), 43–57 and Terry Pinkard, *Hegel's Phenomenology: The Sociality of Reason* (Cambridge, England and New York: Cambridge University Press, 1994), 134–46, 233–49. An influential critique of the alleged rationality of Greek society is found in Jean-Pierre Vernant, *Myth and Society in Ancient Greece*, trans. Janet Lloyd (New York: Zone Books, 1988). See also Hannah Arendt's classic discussion of the strict separation between the Greek polis and the household in *The Human Condition* (Chicago: The University of Chicago Press, 1958), 28–37.

21. For a detailed discussion of the philosophical and social implications of abstract exchange, see Alfred Sohn-Rethel, *Intellectual and Manual Labor: A Critique of Epistemology*, trans. Martin Sohn-Rethel (Atlantic Highlands, N. J.: Humanities Press International, 1983). Written in 1951, Sohn-Rethel's book anticipates Lefebvre's notion of abstract space, as should be clear in such passages as the following: "Time and space rendered *abstract* under the impact of commodity exchange are marked by homogeneity, continuity and emptiness of all natural and material content, visible or invisible (e.g., air). The exchange abstraction includes everything that makes up history, human and even natural history. The entire empirical reality of facts, events and description by which one moment and locality of time and space is distinguished by another is wiped out. *Time and space assume thereby that character of absolute historical timelessness and universality which must mark the exchange abstraction as a whole and each of its features.*" Italics added.

22. "Modernity is doomed to explore and live through abstraction. Abstraction is a bitter chalice, but modernity must drain it to the dregs and, reeling in simulated inebriation, proclaim it the ambrosia of the gods. Abstraction perceived as something concrete, antinature and a growing nostalgia for nature which has somehow been mislaid—such is the conflict lived out by 'modern' man." Lefebvre, *Introduction to Modernity: Twelve Preludes: September 1959–May 1961*, 193. It should be obvious from this passage that Lefebvre's engagement with the problem of abstraction and modernity predates *The Production of Space*.

23. "The abstraction which renders embodied labour abstract labour is a social abstraction, a real social process quite specific to capitalism. Abstract labour is not a way of reducing heterogeneous labours to the common dimension of time, via the commodity relations of the labour process, but has a real existence in the reality of exchange . . . And it is only in the exchange process that heterogeneous concrete labours are rendered abstract and homogeneous, that private labour is revealed as social labour. It is the market which does this; and so there can be no **a priori** determination of abstract labour. Colletti goes further and argues that not only does the abstraction emerge out of the reality of exchange, but also that abstract labour is alienated labour: exchange provides the moment of social unity in the form of an abstract equalization or reification of labour power in which human subjectivity is expropriated." Entry on "Abstract Labour" in Tom Bottomore, ed., *A Dictionary of Marxist Thought* (Cambridge: Harvard University Press, 1983), 2. Lefebvre's understanding of abstract space is compatible with this account of abstract labor, especially its emphasis on exchange as a mechanism of reification and the expropriation of subjectivity. For a recent attempt at recon-

structing the notion of abstract labor, see Moishe Postone, *Time, Labor and Social Domination* (Cambridge: Cambridge University Press, 1996).

24. For a discussion of incipient abstraction and commodification in the eighteenth century, see Christoph Asendorf, *Batteries of Life: On the History of Things and their Perception in Modernity*, trans. Don Reneau (Berkeley and Los Angeles: University of California Press, 1993), 2–3.

25. Guy Debord, *The Society of the Spectacle*, trans. Donald Nicholson-Smith (New York: Zone Books, 1994), 12.

26. Debord, *The Society of the Spectacle*, thesis 165, page 120. It is, of course, equally plausible that Debord's usage of abstract space was inspired directly or indirectly by Lefebvre. The film version of Debord's book appeared in 1973, and its images of massive French housing and urban redevelopment projects provide vivid illustrations of what Lefebvre calls abstract space. Although mentioned only twice in *The Production of Space*, Debord in my judgment occupies a greater prominence in the text than this scant acknowledgment would suggest. A thorough inventory of his influence upon Lefebvre's work is beyond the scope of this article, but in addition to the probable origin of the phrase abstract space in *The Society of the Spectacle*, I would note four other significant areas of overlap between the two books. First, in his claim that the spectacle "unites what is separate" (thesis 29/page 22), Debord employs a Hegelian vocabulary of universalization and particularization that presents the function of the spectacle as similar to that which Lefebvre attributes to abstract space. (Anselm Jappe explores Debord's Hegelianism in his recent study *Guy Debord* (Marseilles: Via Valeriano, 1996), although here *The Phenomenology of Spirit* and its account historical and temporal *Aufhebung* appears the most decisive intertext.) Second, the discussion of spatialization of time and the alienation of temporal and spatial experience (thesis 31/page 23, thesis 161/page 116) coincides with Lefebvre's definition of abstract space. Third, Debord's assertion that "everything that had formerly been absolute became historical" (thesis 73/page 48) prefigures Lefebvre's account of the transformation between these two spatial modes. Finally, both the notions of spectacle and abstract space share a strong anti-ocularcentric bias. For discussions of the relationship between Debord and Lefebvre, see Lefebvre, *le temps de méprises* (Paris: Stock, 1975) and Kristin Ross, "Lefebvre on the Situationists: An Interview," *October* 79 (Winter 1997): 69–83.

27. Lefebvre traces his interest in urban space to his experience watching the construction in 1953–54 of the abstract space of "nouvelle ville" in the oil town of Lacq-Mourenx. See Ross, "Lefebvre on the Situationists," 76.

28. Georg Lukács, *History and Class Consciousness*, trans. Rodney Livingstone (Cambridge; The MIT Press, 1971), esp. 83–110. For a critique of the notion's usefulness, see Hanna Fenichel Pitkin, "Rethinking Reification," *Theory and Society* 16 (1987): 263–93.

29. "A special quality may also be provided by one or several relatively narrow ranges of frequencies in which the coupling from the instrument to the surrounding air is stronger than at other frequencies. These ranges in frequency are referred to as formants. In the human voice, the several formant frequencies are varied by adjusting the shape of the vocal tract, and in speaking, the distinction among the

various vowel sounds is made by appropriate subconscious adjustment or tuning of the various formant frequencies. In the singing voice these same adjustments are heard not only as differences among vowel sounds but also as changes in vocal timbre or tone quality." Entry on "acoustics" in Don Michael Randel, ed., *The New Harvard Dictionary of Music* (Cambridge: Harvard University Press, 1986), 10–11. We might understand Lefebvre's appropriation of the notion of formant to signify those discourses in which the ideological valence of space is most pronounced. Interpreted in this manner, a formant would comprise a cultural, social, or political domain where spatial contradictions and tensions are generated.

30. Neil Smith astutely suggests that Newtonian absolute, Cartesian conceptual, and Kantian a priori space provide the philosophical bases for Lefebvre's theory of abstract space. "ASN." page 49 this volume.

31. Martin Heidegger, "The Age of the World Picture" in *The Question Concerning Technology*, trans. William Lovitt (New York: Harper and Row, 1977), 142; Lefebvre, (*POS*, 289). One can also discern echoes of Heidegger's distinction in *Being and Time* (1927) between what is merely present (*vorhanden*) and what is at hand (*zuhanden*) in Lefebvre's discussion of the problem of defining abstract space as what is unproblematically given and objectified in seeing. For a helpful treatment of Heidegger on vision, see Stephen Houlgate, "Vision, Reflection, Openness: The 'Hegemony of Vision' from a Hegelian Point of View" in *Modernity and the Hegemony of Vision*, ed. David Michael Levin (Berkeley and Los Angeles: University of California Press, 1993), 87–123.

32. The Marxist theory of the concrete totality employed in *The Production of Space* was previously adumbrated by Lefebvre in his 1940 book *Dialectical Materialism*, trans. John Sturrock (London: Jonathan Cape, 1968). The notion of dialectical logic appears in his *logique formelle logique dialectique* [1947] (Paris: Editions Anthropos, 1969), esp. 59–103 and 147–227. For a discussion of Lefebvre's relationship to Lukács and the Hegelian understandings of totality in Western Marxism, see Martin Jay, *Marxism and Totality: The Adventures of a Concept: From Lukács to Habermas* (Berkeley and Los Angeles: University of California Press, 1984), 295. Lefebvre discusses his own intellectual debt to Hegel briefly in "An Interview with Henri Lefebvre," *Environment and Planning D: Society and Space* 5 (1987): 34 and provides his criticisms of Lukács in *la somme et le reste* (Paris: Belibaste, 1973), 179–184. For a divergent perspective see Andy Merrifield, "Lefebvre, Anti-Logos, and Nietzsche: An Alternative Reading of *The Production of Space*," *Antipode* 27:3 (1995): 294–303. I find Merrifield's argument for Lefebvre as a Nietzschean thinker unconvincing and informed by an overly simplistic understanding of Hegel and the dialectic. A recent and more subtle account of dialectical thinking is found in David Harvey, *Justice, Nature, and the Geography of Difference* (Oxford and Cambridge, Mass.: Basil Blackwell, 1996), 46–68.

33. For an intriguing discussion of Hegel's *Philosophy of Right* in relation to the theory of imperialism, see David Harvey, "The Spatial Fix: Hegel, Von Thunen, and Marx," *Antipode* 13 (1981): 1–12.

34. "Abstract space reveals its oppressive and repressive capacities in relation to time. It rejects time as an abstraction—except when it concerns work, the producer of things and surplus value. Time is reduced to constraints of space:

schedules, runs, crossings, loads." Lefebvre, "Space: Social Product and Use Value," 287.

35. A notable exception is the work of Derek Gregory, one of the few commentators who has not shied away from the Hegelian and historicist dimensions of Lefebvre's thought or declared them to be politically circumspect *tout court*. See his *Geographical Imaginations*, 359–62, 377. For a critique of this historicist reading of Lefebvre, see Edward Soja, *Third Space: Journeys to Los Angeles and Other Real Imagined Places* (Oxford and Cambridge, Mass.: Basil Blackwell, 1996), 73, 173–83.

36. Georg Wilhelm Friedrich Hegel, *Hegel's Science of Logic*, trans. A.V. Miller (Atlantic Highlands, N.J.: Humanities Press International, 1990), 600–22. A handy German edition of the book is the Suhrkamp edition, Georg Wilhelm Friedrich Hegel, *Werke* 6 *Wissenschaft der Logik* (Frankfurt a.m. Suhrkamp Verlag, 1969), vol. 2: 273–301. Helpful English-language discussions include John W. Burbidge, *On Hegel's Logic: Fragments of a Commentary* (Atlantic Highlands, N. J.: Humanities Press International, 1981), Errol E. Harris, *An Interpretation of the Logic of Hegel* (Lanham, Md.: University Press of America, 1983), and Michael Inwood, *A Hegel Dictionary* (Oxford and Cambridge, Mass.: Basil Blackwell, 1992).

37. Inwood, *A Hegel Dictionary*, 59.

38. Ibid., 304.

39. Harris, *An Interpretation of the Logic of Hegel*, 42.

40. It is instructive to note Lefebvre's anti-idealist assertion of the unity of knowledge and concepts with experience: "When the critical moment occurs, **connaissance** generates the **concrete universal**. The concepts necessary (among them that of **production**) are not sufficient unto themselves: they lead back to the practice they hold up to view" (*POS*, 368).

41. A cogent definition of the concrete totality is found in Lefebvre's earlier work: "The analysis of the given reality, from the point of view of political economy, leads to 'general abstract relations': division of labour, value, money, etc. If we confine ourselves to the analysis we 'volatilize' the concrete determinations, and lose the concrete presupposed by economic categories, which are simply 'abstract, one-sided relations of an already given concrete and living whole'. This whole must be recovered by moving from the abstract to the concrete. The concrete totality is thus the conceptual elaboration of the content grasped in perception and representation: it is not, as Hegel thought, the product of the concept begetting itself above perception and representation." Lefebvre, *Dialectical Materialism*, 87.

42. Karl Marx, *Grundrisse*, trans. Martin Nicolaus (New York: Vintage Books, 1973), 101. See also Hiroshi Uchida, *Marx's Grundrisse and Hegel's Logic*, ed. Terrell Carver (London and New York: Routledge, 1988) and Lucio Colletti, *Marxism and Hegel*, trans. Lawrence Garner (London: NLB, 1973), 113–38.

43. For discussions of the notion of the phantasmagoria, see Terry Castle, "Phantasmgaoria: Spectral Technology and the Metaphorics of Modern Reverie," *Critical Inquiry* 15 (Autumn 1988): 26–61. Siegfried Kracauer described the repetition of the standardized bodies of the Tiller Girl dancers in his famous 1927

essay "The Mass Ornament." See Siegfried Kracauer, *The Mass Ornament*, trans., ed. and introduction by Thomas Levin (Cambridge: Harvard University Press, 1995), 75–86.

44. Debord, *The Society of the Spectacle*, thesis 34/page 24.

45. For an insightful discussion of Lefebvre's "mourning" for the loss of absolute space in relation to the work of Jacques Lacan, see Pile, *The Body and the City*, 166–67.

46. Henri Lefebvre, *The Survival of Capitalism* (1973), trans. Frank Bryant (London: Allison and Busby, 1976), 85.

47. G. W. F. Hegel, *Hegel's Science of Logic*, trans. A. V. Miller (Atlantic Highlands, N. J.: Humanities Press International, 1989), 609.

48. Lefebvre, *The Production of Space*, 306–07, and *Hegel's Science of Logic*, 610.

49. Edward Casey, *The Fate of Place: A Philosophical History* (Berkeley and Los Angeles: University of California Press, 1997).

50. Lefebvre's notion of difference is outlined in *Le manifeste différéntialiste* (Paris: Gallimard, 1970).

51. See David Harvey, *The Condition of Postmodernity* (Oxford and Cambridge, Mass.: Basil Blackwell, 1989), 219. While I am in overall agreement with Harvey's comment that Lefebvre's account of the relation among the spatial concepts is "much too vague," I interpret *The Production of Space* more as an ontological theory of space—and less as a spatial history—than does Harvey.

52. This passage is an echo of Marx's famous remark in the *Grundrisse* (105) about human anatomy providing a key to the anatomy of the ape.

53. Geographers would be quick to point out the structuration of human settlement and social organizations by the physical and natural features of space. Although Lefebvre is not primarily interested in this meaning of spatial interpenetration, his theory does not preclude it.

54. See Alexandre Kojève, *Introduction to the Reading of Hegel: Lectures on the Phenomenology of Spirit*, and assembled by Raymond Queneau, ed. Allan Bloom, trans. James A. Nichols, Jr., (Ithaca, N.Y.: Cornell University Press, 1980).

55. Lefebvre, *The Production of Space*, 55. An interesting challenge to Lefebvre's invocation of class struggle as the only force capable of checking abstract space is Edward S. Casey's suggestion that place itself might be able to resist the force of abstract space. See his "The Production of Space, or The Heterogeneity of Place" this volume page 71.

56. The most recent English translation of Lefebvre's work to appear is *Writings on Cities*, ed. Eleonore Kofman and Elizabeth Lebas (Oxford, England and Cambridge, MA.: Basil Blackwell, 1996), 13.

57. The ability to locate any object on the earth's surface with extreme accuracy provided by Geographic Information Systems (GIS) suggests an astonishing expansion of the realm of abstract space ripe for Lefebvrean analysis. For critical discussions of this technology, see Laura Kurgan, "You are Here: Information Drift," *Assemblage* 25 (1995): 16–43 and John Pickles, ed. *Ground Truth: The Social Implications of Geographic Information Systems* (New York and London: The Guilford Press, 1995).

58. Henri Lefebvre, *Everyday Life in the Modern World* trans. Sacha Rabino-vitch (New York: Harper and Row, 1971). The discussions of the "pure formal space" of terror (179) and the city (190) prefigure the notion of abstract space in their description of a unified space from which time has been evicted.

59. Here we might do well to consider Deutsche's feminist critique of the "foundationalist image of society as a fiction of subjects driven by a desire to disavow their own partial and fragmented condition" that she discerns in much contemporary spatial theory, *Evictions*, xx. I believe it prudent to maintain a healthy suspicion of totalizing generalizations which obfuscate difference (such as "the quality of life" in a city or "the body" of the urbanite in general without reference to the factors of gender, race, class, or age) but less justifiable to accept the poststructuralist critique of history and representation that informs Deut-sche's analysis of contemporary urban space and contravenes the spirit of Lefeb-vre's project. Although Deutsche was one of the earliest cultural critics to employ Lefebvre's notion of abstract space to impressive effect in her 1988 essay "Uneven Development: Public Art in New York City,"(*Evictions*, 49–108), she rejects the universalizing and historicizing dimensions of *The Production of Space* and what she calls Lefebvre's "humanist belief in a previously integrated social space" (77). Her claim that "stories about the beginning of public space are not really about the past; they tell us about the concerns and anxieties inhabiting our present social arrangements" (290) evinces a radical presentism that denies the possibility of any knowledge of past spatial forms that could prove useful in devising contemporary spaces and strategies of resistance and contestation. I think one might still learn from the history of urban and spatial forms and recognize their tendency to har-bor traces of anterior spaces, as Lefebvre clearly does, without adopting his nos-talgic privileging of particular historical spaces. To construe spatiality as merely epiphenomenal to political and social conflict and the politics of representation is to confine understanding of its effects to a single narrowly circumscribed tempo-ral location that forecloses the possibility of relating the particular to the univer-sal, the local to the global, that Lefebvre understands as the production of spatial knowledge. Is it not possible to analyze and make distinctions about different historical modalities of public space without valorizing "a former golden age of total knowledge" or elevating oneself as a scholar to "a position outside the world" (318)? By disregarding those anterior spatial forms that underpin Lefeb-vre's understanding of contemporary abstract space, Deutsche's poststructuralist appropriation of his work distances itself from the central *political* aspirations of *The Production of Space*.

60. See Neil Smith, "ANS," this volume, page 49. See also Edward S. Casey, *Getting Back into Place* (Bloomington: Indiana University Press, 1993), especially chapters 3 and 4 and idem, *The Fate of Place: A Philosophical History*, especially chapter 10.

61. Even Lefebvre's erstwhile student, Guy Debord, was more sanguine than Lefebvre about the prospect of images critically engaging with and exposing the "society of the spectacle." For an account of Debord's understanding of how film might reveal its mechanisms, see Thomas Y. Levin, "Dismantling the Spectacle: The Cinema of Guy Debord" in Elisabeth Sussman, ed., *On the passage of a*

few people through a rather brief moment in time: The Situationist International, 1957–1972 (Cambridge: The MIT Press, 1989), 72–123. The standard discussion of French anti-ocularcentrism is Martin Jay, *Downcast Eyes: The Denigration of Vision in Twentieth-Century French Thought* (Berkeley and Los Angeles: University of California Press, 1993). Jay provides a useful discussion of Lefebvre and Debord in chapter 7. The work of Michel Foucault, especially *Discipline and Punish: The Birth of the Prison* (1975), trans. Alan Sheridan (New York: Pantheon, 1979) resonates throughout *The Production of Space*, but apart from a few dismissive remarks about Foucault's epistemology, Lefebvre offers little commentary on his spatial and anti-ocularcentric views. Although Foucault's book appeared in France a year after Lefebvre's, it is likely that he was familiar with the project from Foucault's highly publicized lectures at the Collège de France.

62. Gregory, *Geographical Imaginations*, 403.

63. Lefebvre, *The Production of Space*, 290, 302. I disagree with Gregory's characterization of *The Production of Space* as historicist (*Geographical Imaginations*, 359–362), for Lefebvre's steadfast refusal to specify the moment of origin of abstract space and to identify a telos of spatial history distinguishes his work from traditional understandings of historicism as an organic unfolding of historical stages. Lefebvre's suggestion that abstract space emerges around the time of the French Revolution is confirmed by Anthony Vidler's argument that the institutions of the hospital, the prison, the asylum, and the workhouse appeared within the period from 1770 to 1789. See Anthony Vidler, *The Writing of the Walls* (Princeton, N.J.: Princeton Architectural Press, 1987), 51.

64. R. J. Horvath and K. D. Gibson, "Abstraction in Marx's Method," *Antipode* 16, no. 1 (1984):19. I have learned much from this stimulating article.

65. Gregory, 382.

Antinomies of Space and Nature in Henri Lefebvre's *The Production of Space*

Neil Smith

"Capitalism," Henri Lefebvre once remarked in a famous declaration, "has found itself able to attenuate (if not resolve) its internal contradictions for a century" since Marx wrote *Capital*. "We cannot calculate at what price," he concluded, "but we do know the means: *by occupying space, by producing a space*."[1] Even by Lefebvre's standards this must have seemed an idiosyncratic claim. An iconoclast of the French left—from his involvement with the Surrealists and the *Philosophes* group in the 1920s, membership in the Resistance, troubled collaboration with the Situationists, and sometime inspiration in the general strike of May 1968—Lefebvre's influence even in France, but especially in anglophone Marxism, has been strangely muffled.[2] His most theoretical contributions were often received as tangential to the political currents of the period, their import deflected as much as absorbed, and outside a committed group of followers he is only begrudgingly acknowledged as a central force in twentieth century Western Marxism. Of an astonishing sixty-six books, many widely available in German, Spanish, and other languages, only five were, until the 1990s, translated into English.

Yet the translation of two of Lefebvre's most important books into English, the much heralded advent of a so-called "spatial turn," and probably also Lefebvre's death in June 1991 in his native Pyrenean foothills, at the age of 90, have all combined to inspire something of a Lefebvre revival, at least in English language intellectual circles.[3] It is appropriate if serendipitous that the two recently translated books—*The Critique of Everyday Life* and *The Production of Space*—appeared concurrently in English in the 1990s because although they were originally published twenty-seven years apart, Lefebvre, by the early 1970s, was coming to the conclusion that everyday life, with which he had been concerned since the 1920s, could only be demystified through an understanding of what he refereed to as "the production of space." As he put it elsewhere, "time has disappeared in the social space of modernity."[4]

This renewal of interest in Lefebvre highlights his historical contributions to Western Marxism and social theory more broadly, but it also contributes to the contemporary transformation of social and political theory. This essay examines several central arguments in Lefebvre, focusing on a series of critical antinomies that display both some strengths and some weaknesses in Lefebvre.

I do not think it is necessary to rehearse yet again the history of the spatial turn within which the revival of interest in Lefebvre is occurring.[5] But I do think it is worth pointing out, in the spirit of Lefbvre's own project of physical, mental, and social space, that the timing of this rediscovery of space, in which he himself participated, is significant. Emerging from the 1960s, it coincides broadly with the dissolution of a comparatively stable political, cultural and economic geography of postwar expansion. At the global scale, the "American Century," as Henry Luce dubbed it in 1942, was coming to an end after barely three decades; the traditional division between the so-called First and Third Worlds fragmented as newly industrializing economies differentiated themselves from the increasingly redlined economies of sub-Saharan Africa; and by 1989 the Second World ceased to exist entirely, as such.[6] At the national scale, decolonization throughout Africa and resurgent nationalisms from Scotland to the Balkans have either hardened or redrawn the geopolitical boundaries of the nation state while the halting solidification of the European Community and the growing porosity of national boundaries vis-à-vis international capital have shattered the brief geographical stability of the postwar system of nation states. At the regional scale developed regions have experienced dramatic deindustrialization while new regional ensembles, devoted to post-Fordist production and service provision, have emerged from Silicon Valley to Taipei to the M4 corridor of Southern England. At the urban level too, the traditional division in the developed world between a declining central city and expanding, functionally segregated suburbs no longer holds true. From the redevelopment and gentrification of central cities to the functional (but not social) integration of the suburbs, urban space too is being systematically restructured. In the underdeveloped world, incomplete integration into the world economy has exacerbated the intense spatial dichotomy between the metropolitan centre and the working class periphery.

In short, the solidity of the geography of twentieth century capitalism at various scales has melted; habitual spatial assumptions about the world have evaporated. We can add to this the equally dramatic problematization of the body as political reconstructions of race, gender, class, and sexual-

ity during the same period rework the cultural economy of identity. It is
as if the world map as jig-saw puzzle had been tossed in the air these last
two decades, leaving us to reconstruct a viable map of everything from
bodily and local change to global identity. Under these circumstances, the
taken-for-grantedness of space is impossible to sustain. Space is increasingly revealed as a richly political and social product, and putting the jig-
saw puzzle back together—in practice as well as theory—is already a
highly contested affair. Bush's "new world order" was a nostalgic return
to 1945, a second try at Yalta, and it attempted an appropriately spatial
vision of global domination cloaked in predictable moral pieties. NAFTA
and stumbling efforts at "European integration" are also part of the efforts to reconstruct the geography of capitalism along predictably neo-
liberal lines. Another more prosaic vision comes from the U.S. Congress
which, in 1987, passed a joint resolution establishing an annual "Geography Awareness Week," designed to enhance geography teaching "in the
national interest." In Britain, in the early 1990s, the high school geography curriculum became a political battleground in a way that could not
have been imagined two decades earlier. And at the beginning of the
1990s, Casper Weinberger, Reagan's Secretary of Defense turned pub-
lisher, issued an editorial plea in his *Forbes* magazine (self-described
"Capitalist Tool") to "Bring back geography" in U.S. schools and univer-
sities. This was vital, he insisted, because of the "heartening daily rejection
of Communism" and the general upheaval in world events. "We can profit
in every sense of the word from these amazing changes," he concluded,
but "all of this starts with geography."[7]

Henri Lefebvre was an early and ardent proponent of a respatialized
social theory. His proposal that we think and act politically in terms of
the "production of space" was as iconoclastic as his analysis was suggestive, and yet its insights and especially its critiques have had only a glanc-
ing political effect in France and are only beginning to be considered in
the anglophonic world. If Lefebvre offers some cautionary alternatives to
the dense verbal spatialization of much French social theory, it is a cau-
tion that is even more relevant today with the translation and political
transubstantiation of much of this work into the academic core of the
humanities in the anglophonic world. In particular, his focus on the social
production of space is a useful antedote to the increasingly conservative
idealism of the narrowest proclamations of social constructionism. Further, Lefebvre's spatialized politics point toward the possibility of an "integral marxism" capable of joining the struggle to reconstruct the globe
out of the territorial and ideological fragments now at hand.

And yet there are tenacious problems. In the first place, an unresolved
contradiction between ontology and history drives much of Lefebvre's
vision. This is undoubtedly reaffirmed by a sometimes quixotic resort to

a philosophism that he himself, in other contexts, would critique. And it leads to an uncharacteristically simple dismissal of an unreconstructed nature. Lefebvre's socialization of space is as incomplete, therefore as it is impressive. But second, although I am myself drawn to it, I am not entirely convinced by Lefebvre's explanation for the necessity of spatializing social theory. Why, to put it most simply, is politics now a question of space? Third, while the intractable (and largely unedited) language of his exposition may enhance his deconstructive intent, Lefebvre gives us surprisingly little sense of what alternative productions of space might look like—how an alternative spatialized politics would proceed. He aspires toward and in part predicts a "redifferentiation" of space, but what does this mean in practice? I will take up the question of geographical scale, in the context of the material and ideological reconstruction of space, in an effort to push a spatialized politics toward better definition.

The Critique of Idealism and the Production of Space

"If the search for a unitary theory of physical, mental and social space was adumbrated several decades ago," Lefebvre observes, "why and how was it abandoned?"(p. 21).[8] In *The Production of Space*, written in 1973, Lefebvre seeks to re-initiate the search for this unitary theory of space. He has in mind as precursors of this unified theory the Surrealists, and various technological utopias, none of which came to social fruition. Such a "science of space" is a sine qua non of any successful anti-capitalist politics, he argues. His first target, then, is the conceptual history and the social actors who have brought us to the contemporary fragmentation, abstraction, and depoliticization of space. In philosophical terms, Descartes's announcement of space as absolute, homogeneous, an infinite *res extensa* marks simultaneously the decisive shift raising the possibility of a "science of space," and yet at the same time invites its demise. Thereafter excised by mathematicians, he suggests, space was reconceptualized in increasingly abstract and multiple ways: non-Euclidean space, n-dimensional space, curved space, and the like all encouraged the definition of space in terms of mental rather than physical phenomena. Space was eviscerated—before anything else it was rendered a conceptual abstraction. "No limits at all have been set on the generalization of the concept of *mental space*," he complains, not just in philosophical but in social discourse. But space that remains "mental space" or "theoretical space" is no space at all: "the philosophico-epistemological notion of space is fetishized and the mental realm comes to envelop the social and physical ones. . . . Conspicuous by its absence from supposedly fundamental epistemological studies is not only the idea of 'man' but also that of space—

the fact that 'space' is mentioned on every page notwithstanding" (pp. 3, 5). The assumed identity between "mental space" and "real space" creates an unbridgeable abyss that evades social practice and physical change.

Apt a quarter century ago, Lefebvre's critique of the "spatial turn" in postwar French social theory and philosophy should have special poignancy today. He anticipates some of the dangers of a partial spatialization that remains narrowly mental, resisting the empirics of physical or social space, and anticipates too how such a move greases the wheels for a rightward moving idealism. "What is happening here," he argues, "is that a powerful ideological tendency . . . is expressing in an admirably unconscious manner, those dominant ideas which are perforce the ideas of the dominant class. . . . When codes worked up from literary texts are applied to spaces—to urban space, say—we remain . . . on the descriptive level, . . . This is to evade both history and practice" (pp. 6,7).

Lefebre has in mind here Foucault and Althusser especially, but also Sartre, Derrida, Chomsky, and Kristeva, for whom the adoption of a spatial grammar was not only opportunistic, he argued, but politically charged. He reserves his most devastating condemnation for them:

> Most if not all authors ensconce themselves comfortably enough within the terms of mental (and therefore neo-Kantian and neo-Cartesian) space, thereby demonstrating that "theoretical practice" is already nothing more than the egocentric thinking of specialized Western intellectuals—and indeed may soon be nothing more than an entirely separated, schizoid consciousness (p. 24).

Lefebvre clearly anticipates the evolution and adoption of much French social theory into the kind of narrow, increasingly trivial and ideological social constructionism that has become so fashionable today—narrow, trivial, and ideological insofar as it remains within the confines of mental space, trespassing beyond only via the assumption that significations in mental spaces are simultaneously transformations of physical and social space. This assumption of an unproblematic mental mapping perpetuates rather than challenges the ruling hegemonies, Lefebvre argues. "The aim of this book," he announces, "is to detonate this state of affairs" (p. 24).

The central question remains, then: How to reunite this disembodied "ideal" space with "real" space (p. 14)? How are *physical* space ("nature, the Cosmos"), *mental* space ("including logical and formal abstractions") and *social* space to be recombined (p. 11)? Lefebvre's answer lies in reappropriating Marx's "concept of *production* and of the *act of production*" as concrete universals in Hegel's sense. Through definite social acts and practices, he argues, physical, mental, and social space are produced, reproduced, and transformed in constant connection, and whatever the extraordinary achievements of the social imagination, only by subsequent abstraction can mental space flatter itself concerning its independence from these social practices. "To speak of 'producing space' sounds bizarre, so great is the sway still held by the idea that empty space is prior to whatever ends up filling it" (p. 15). For Lefebvre it is impossible to retain a conception of space as external to human activity, pre-given, an a priori field across which social and natural activity takes place. Space is in any meaningful sense produced in and through human activity and the reproduction of social relations.

Different societies and different modes of production produce space differently; they produce their own kinds of spaces. In Lefebvre's language, specific societies and specific periods have distinctive spatial codes that may to some extent be known, read, decoded. Such codes are integral to the social and spatial practices of a given place and period, and the job of theory is to elucidate the emergence, performance, and collapse of these spatial codes.

A triad of elements contribute to any particular production of space, and insofar as Lefebvre's spatialization of social theory is a quintessentially political project, any particular comprehension of the production of space involves an analysis of these elements. He seeks to decode *representational spaces*—socially produced spaces imbued with more or less structured social meanings that are directly lived—and to critique specific conceptualizations or *representations of space*, but above all he seeks to inform and develop critical *spatial practices* via which social space is perceived and via which social difference could be democratically re-asserted (p. 38–39, 46). The production of space is not just the powerful accomplishment of capital and the ruling classes but the ambition as well as means of liberatory reconstruction of the social world. With the "disappearance of time in the social space of modernity," it is space not time, Lefebvre insists, which masks the configurations of social power.

But Why Space? Ontology and History

Lefebvre proposes what he calls a "strategic hypothesis," which is simultaneously ambitious and yet understated:

> Theoretical and practical questions relating to space are becoming more and more important. These questions, though they do not suppress them, tend to resituate concepts and problems having to do with biological reproduction, and with the production both of the means of production themselves and of consumer goods (p. 62).

But why space? Why does Lefebvre opt for a thorough reconstruction of politics—the unstated implication of this hypothesis—as a spatial rather than historical practice? The historical privileging of time over space is indisputable, but does not a social politics premised in and through the "production of space" bend the stick back too far? What happens to history in this vision?

There are always in Lefebvre two kinds of answers to this question. The one historical and the other ontological. The ontological answer emerges from Lefebvre's encomium for Hegel. Hegel, he proclaims, "is a sort of Place de l'Etoile with a monument to politics and philosophy at its centre. According to Hegelianism," he continues, "historical time gives birth to that space which the state occupies and rules over" (p. 21). Marx's reinstatement of revolutionary time, Bergson's temporal immediacy in consciousness, Husserl's phenomenology, and Lukács' imprisonment of space as (in turn) the jailer of false consciousness, all represented in very different ways a "restoration of time" against Hegel's dramatic acclamation of space—his "fetishization of space in the service of the state" (p. 21–22). Lefebvre appreciated Nietzsche for resisting this philosophical back-tow, even if his embrace of the simple positivism and dramatic force of absolute space nonetheless contrasted dramatically with Hegel.

In one sense, we might say that Lefebvre now reciprocates by doing for space in Marxist theory, what Lukács' *History and Class Consciousness* did for time. *The Production of Space* is fundamentally about the spatiality of everyday life and its abstractions, the material *cum* conceptual geography of local and global existence, and the succession of spatial regimens that has carried us to the political present. Both Lukács and Lefebvre strain their Marxism through Hegel, but while Lukács grasps the temporal determination of social change, Lefebvre leans to Hegel's "end of history"[9] and takes his stand there, on "space" as the "product and residue of historical time." History "disappears" and time comes to be "dominated by repetition and circularity, overwhelmed by the establishment of an immobile space which is the locus and environment of all Reason," and thereby "loses all meaning"(p. 22). Geography in the broadest sense, not history, is the crucial result of this Hegelian movement, for Lefebvre.

But insofar as mental space is never dissociated from social space, this philosophical victory of space—albeit always challenged and inevitably incomplete—is paralleled by an historical explanation for the pre-eminence of space. The second half of the twentieth century has witnessed three vital shifts. First, "the state is consolidating on a world scale," imposing increasingly integrated and burdensome forms of knowledge, technology, and ideology on different societies and places:

Space in its Hegelian form comes back into its own. The modern state promotes and imposes itself as the stable centre—definitively—of (national) societies and spaces. As both the end and the meaning of history—just as Hegel had forecast—it flattens the social and "cultural" spheres (p. 23).

But, second, the wars, revolutions, victories, and defeats of the period represent an inevitable opposition to the imposed rationality of the state, a new negativity that "corresponds precisely to Nietzsche's vision."[10] Yet third, "nor has the working class said its last word" (p. 23).

All of this, Lefebvre argues, is premised on "an epoch-making event" that is "generally ignored," namely the smashing of the traditional spatial codes that had dominated Western societies since the sixteenth century: "The fact is that around 1910 a certain space was shattered. It was the space of common sense, of knowledge (*savoir*), of social practice, of political power," the space alike of Euclidean geometry, Renaissance perspectivism, and descriptive geographies. The artistic revolution initiated by cubism announced most vividly the advent of the new space, and Picasso's 1907 *Le Demoisseles d'Avignon* marks this shift precisely. "Picasso's space *heralded* the space of modernity." He "glimpsed the coming dialectical transformation of space and prepared the ground for it . . . , discovering and disclosing the contradictions of a fragmented space" (pp. 25, 300–4).[11]

Now, although I am inexorably drawn to Lefebvre's spatial vision and convinced by the brilliance of his proposal of the "production of space," nonetheless, I do not find this simultaneous ontological and historical justification for recentering space to be entirely convincing. On the one hand, he presents Hegel's ontology, which he struggles to strip of its barest teleology; on the other, he presents Picasso and the global consolidation of the state—shards of an empirical history. The connection between the two is far more than co-incidence for Lefebvre—Hegelian space "comes back into its own" with the events of the twentieth century—yet less than complete assimilation. Ontology and history are correlated but not in some simple one-to-one mapping; they are, we might say, "malcongruous." This abiding malcongruity of ontology and history might itself be seen as symptomatic of the current evisceration of space, but surely only by positing an even more problematic affinity of space and ontology that risks simply reversing the privileging of time with that of space.[12]

Lefebvre's is an anti-philosophical philosophy. His beginning discussion is rooted deep in philosophy as a means of criticizing the isolated abstractions and fetishized concepts of space that "philosophico-epistemological" thought has produced. And the converse might be said of his history: it is an historical anti-history. Insofar as ontology and history struggle with each other throughout the text, therefore, his philosophy reaffirms his anti-history, and vice versa, while his history reaffirms the anti-philosophy.

This is perhaps clearest in Lefebvre's history of space. The history of society is indeed the story of class struggle, for Lefebvre, but of class struggle *for space*. For the official Marxist periodization of primitive communism, slavery, feudalism, and capitalism, followed by socialism, he substitutes an alternative and more abstract schema that confirms that the history of social life *is* the history of space, socially produced: absolute space, historical space, abstract space, and differential space succeed (if not entirely supersede) one another.

For Lefebvre, *absolute space* "was made up of fragments of nature," settled and socialized for various reasons of environment, location, and resources. It was "natural space" populated, space in which direction and dimension are not abstracted but expressed in terms that fuse practicality and symbolism. Peculiarly Western notions of absolute space, conveying a sense of empty, measured, homogeneous volume derive from Greek conceptual orders (pp. 236–37). The progressive differentiation of this space from nature and its internal differentiations lead ultimately to its politicization and the emergence of *historical space*. Usually associated with the growth of cities, historical space accomplished a separation between production and reproduction and is marked by other social differences according to function (division of labour), gender, and territory as well as the origins of property and economic accumulation. Historical space is politicized space. *Abstract space* begins to emerge when social labour is differentiated into an abstract form and social wealth thereby aspires to the possibility of infinite mobility, global hegemony. The imposition of abstract space, in which social difference is continually flattened and "crushed" through the state as well as the economy, is for Lefebvre the hallmark of capitalism. Representations of space gain disproportionate power over spatial practices and spatial representations. Included in abstract space is the space that the populations of most Western societies take for granted, the space of infinite extension ("deep space") and emptiness, the "space" implied by the habitual reference to events happening "in space." "Natural" and instinctive as this notion of space seems, it is an historically and geographically discrete representation of space, however difficult it may be to conceive alternatives. The space of Descartes and Newton became the space of capital, and vice versa; as Lynn Stewart summarizes it, abstract space for Lefebvre is "the codified logic of modern power."[13]

Yet abstract space too is contradictory and tends toward its own redifferentiation and the pupation of *differential space*. The "right to difference" is the most revolutionary claim against the crush of abstract space, for Lefebvre; it "may be achieved through practical action." "The right to difference" is "diametrically opposed to the right of property" and "implies no entitlements that do not have to be bitterly fought for" (p. 396).

If this movement from absolute through historical to abstract space represents a definite historical evolution for Lefebvre, the optimistic "logic" of a subsequent differential space seems to invite a particularly teleological side of Hegel into the history; the philosophy has of course been in the history all along. Not that the Hegelianism of the text is entirely negative. Just as with Hegel, the suspension of one space, subsumed under a subsequent space, in no way erases the pre-existing space. The remnants of absolute space and historical space remain very much alive if transformed in abstract space to the point where they can provoke the disruptive differentiation of abstract space. Intimately connected to constructions of social identity, and constituting the archetypal production of absolute space, for example, the body for Lefebvre is the site of crucial political contest. "Any revolutionary 'project' today," he argued at the dawn of the modern feminist movement, "must, if it is to avoid hopeless banality, make the reappropriation of the body, in association with the reappropriation of space, into a non-negotiable part of its agenda" (p. 166–67).[14]

A spatialized political language provides Lefebvre with a supple means for conceiving universal social structures and local difference simultaneously: the "universality" of capital and its ambition for abstract space as well as the liberatory ambition for a politics based on "the right to difference"; and the balkanization of global experiences according to national boundaries, private property, social differences; and a perverse cultural individualism that simultaneously flattened and balkanized mental space and symbolic possibility. (Here we see why Lefebvre goes so far as to accuse theorists whose appropriation of space remains rigorously mental of ideological collaboration [see quote from p. 24 above]: they participate in the abstraction of space.) A spatialized politics also returned an assortment of excluded questions to the Marxist agenda and introduced a miscellany of others. Lefebvre embraced class struggle but the politics of the production of space open up no less clearly to an analysis of interconnected forms of oppression and experiences of everyday life—the possibility of an integral Marxism.

And yet the weakness of Lefebvre's justification for his own "spatial turn," however markedly it contrasts with the substance of his deliberations on the production of space, provides a questionable grounding for these arguments. The weakness is intensified when we broaden the focus and consider Lefebvre's treatment of nature. This change of focus might at the same time provide a more sustainable framework within which to conceive the production of space.

The Production of Nature?

Lefebvre's rethinking of nature is poor—far less original than his radical repoliticization of space which, in fact, carries along his conceptualization of nature. He leaves nature largely unreconstructed, and with it the relationship between space and nature.

If nature and space co-mingle in absolute space, "primary nature" recedes to the extent that space is historicized; the production of historical space "calls for the immediate production or creation of something other than nature: a second, different or new nature" (p. 109). This disappearance of nature is more actively accomplished with the abstraction of space and the inherent violence that this abstraction implies. Nature is destroyed, pulverized by the expansion of capital, "murdered by 'anti-nature'" (p. 71). It is increasingly left behind, Lefebvre says (thinking of Heidegger), increasingly distant, even "becoming lost to thought" (pp. 121, 31).

And yet, with Hegel, nothing is ever completely lost and precedents continually re-present themselves as novelties. Primary nature may persist in second nature "albeit in a completely acquired and false way . . .—witness urban reality" (p. 229). Likewise, absolute space persists within the relative, and so it is "an irreducible fact" that the "ultimate foundation" of spatial qualities lies in nature; the raw material of production still emanates directly or (increasingly) indirectly from nature. Yet nature is nonetheless squeezed into the background; "a 'second nature' may replace the first, standing in for it or superimposing itself upon it without wreaking complete destruction" (p. 348).

Nature for Lefebvre is on the verge of becoming a corpse at the behest of abstract space. Unlike space, nature retains virtually no initiative to and for itself. "Displaced and supplanted" by social space (pp. 123, 349), it is destroyed on the one hand, and at best "merely reproduced" on the other (p. 376). It is space that progresses and develops, is produced and conjures, space that is alive. Space can even lie, Lefebvre contends, but nature cannot: "A rock on a mountainside, a cloud, a blue sky, a bird on a tree—none of these of course can be said to lie" (p. 81). By contrast, "spaces sometimes lie just as things [commodities] lie":

> a space that is apparently "neutral," "objective," fixed, transparent, innocent or indifferent implies more than the convenient establishment of an inoperative system of knowledge, more than an error that can be avoided by invoking the "environment," ecology, nature and anti-nature, culture, and so forth. Rather it is a whole set of errors, a complex of illusions, which can even cause us to forget completely that there is a total subject which acts continually to maintain and reproduce its own conditions of existence,

namely the state (along with its foundation in specific social classes and fractions of classes). . . . *The space of a (social) order is hidden in the order of space* (pp. 92, 94, 289).

This bifurcated treatment of space and nature emanates ultimately from Lefebvre's handling of production. Production, he argues, is a quintessentially social act, and while nature creates and thereby provides use values for production, it does not produce (p. 68–70). Space and commodities, however are social products, they lie, and social space itself can produce. Now this gives a surprisingly undialectical negativity to the treatment of nature, especially vis-à-vis the life accorded to space. Where the dialectic of space is defined by its political vitality—social and political practice *is* spatial practice—the dialectic of nature[15] is evacuated. Nature is reduced to little more than a substratum.

Lefebvre's analysis is passingly sympathetic with a socialist environmentalism; capitalism via the relentless abstraction of space destroys nature, while the romantic nature-worship of much radical ecology he recognizes as a distraction. He even adopts the Frankfurt School argument that it is the "domination" of nature and not simply its appropriation, as Marx usually put it, that we witness today.[16] But there is little hint of the more active political initiative that the Frankfurt School attributes to nature via the "revenge of nature," or, in place of that admittedly nostalgic construction, some other sense of a politics of nature. Quite the opposite. Nature is increasingly the raw material of mimetic productions of space in which "what are produced are the *signs* of nature or of the natural realm"; "nature is left, as it were, in a no-man's land" (p. 376).

More troublesome, given his appeal to science in constructing a unified theory of space, there is no apparent recognition in Lefebvre of the way in which relativity theory potentially revolutionized the relationship between nature and space. Whereas the notion of "the production of space" brilliantly theorizes the relativity and relationality of social space, it does so at the cost of returning nature to a pre-Einsteinian state. This is especially surprising given Lefebvre's evident appreciation of Leibniz whose "relational space," insisting against Descartes that space was nothing if not occupied, provides a partial precursor to the kind of reconnection of space and matter glimpsed in relativity theory.

In the Newtonian world that relativity theory overturned, space was for the first time radically separated from the material world of nature. In their abstractness, Newtonian (absolute), Cartesian (conceptual) and Kantian (a priori) space (spanning the philosophical basis of Lefebvre's abstract space) all established an unprecedented priority for space over nature; social and natural events took place in a world already delineated in terms of space and time. Beginning most clearly with Ernst Mach at the end of the nineteenth century, physics sought a dual transformation: the reintegration of space and time with nature, which Einstein's relativity theory came to offer; and the reversed priority of space and nature. "The shape of space" was no longer given but rather the product of material events, natural movement; space again could be envisioned as nothing if not occupied. However, this philosophical restructuring of the physical world, and especially the reversal of priority between space and matter, has proven elusive. Anomalies were neutralized insofar as physical space was reinvented in even more abstract forms of mathematical space, and Einstein eventually retreated to the conservative rationalization that the epistemological priority of nature did not imply an ontological priority.

Lefebvre's social "production of space" synchronizes precisely with the intent of relativity theory, then: space is indeed produced. But it does not go far enough. It accomplishes the first transformation but leaves intact the pre-Einsteinian priority of space over nature, a priority which, insofar as physical scientists have not been able to consummate it, may have to await a dramatic socialization of space of the sort that Lefebvre's "unified theory" surely anticipated. He understands, as the Frankfurt School did not, that twentieth century capitalism takes us beyond a dualism of first and second natures—the given and the made—but he remains sufficiently wedded to the externality of nature that he sees only the destruction of nature and not the *production of nature*.

Lefebvre even entertains this possibility only to dismiss it. In introducing the question of social space, he ventures: " 'Nature' itself, as apprehended in social life by the sense organs, has been modified and therefore in a sense produced" (p. 68). The "production of nature" may sound as bizarre and quixotic as the production of space, but if we accept the reversal of priority adumbrated by relativity theory but not delivered on, then this argument becomes of the greatest importance. It does not in any way deny the production of space but suggests, rather, that the production of space is integral to a production of nature and fashioned within it. It is the corollary of the production of nature rather than the other way round.

The argument for the production of nature has been elaborated in broadly parallel terms to Lefebvre's treatment of space.[17] In place of Lefebvre's sequence of absolute, historical, and abstract space, we can appropriate Marx's corollary evolution of production in general, production for exchange, and capitalist production.[18] A second nature is indeed produced out of first nature much as Cicero first argued, and Hegel and Marx reiterated, but this occurs with production for exchange; the production of second nature coincides with the production of historical space. With the advent of capitalist production, this second nature is generalized and with it, of course, the abstraction of space, but not before a dramatic *Aufhebung* takes place. First nature does not so much "remain" in the interstices of second nature as Lefebvre and the Frankfurt School generally propose. Rather, first nature is produced and reproduced from *within* this generalized second nature and as an adjunct to it.

Just like the production of space, the production of nature implies a much more profound shift than the trivial social constructionism that treats nature as concepts *of* nature and necessarily social.[19] At the same time, the fact that the production of nature is a physical as well as a mental process, socially accomplished, in no way implies a regression to the nostalgic romanticism of the "end of nature" or the "death of nature." Nonhuman processes, from gravity to geochemical interaction, atomic fission to biological mutation, are real enough and "natural" enough, but the world in which they operate and their effects are so inextricably bound up with social processes and forms that it is only at considerable cost that science can maintain the fiction that such processes and events can be apprehended, unproblematically, as independent, purely natural forces. This assumption of an independent and available reality is precisely that: an assumption. It is highly useful in many situations, but as a universal philosophy it is absurd. At the level of everyday life, the form of nature is irreparably produced by social activity. First nature is virtually inaccessible, not so much because of the inevitable intervention of language and conceptualization in apprehending it, but because of the ongoing history of physical intervention that has already socialized—in beautiful or grotesque form—what there is to be apprehended.

In this sense we might well reply to Lefebvre that the nature of a social order is also hidden in the order of the nature it produces. And we might add that if space can lie, why not nature too; and lie as well as space in exactly the way that Lefebvre suggests? Cannot a blue sky be a lie on the day of Hiroshima; a cloud lie when it drops acid rain; or a bird lie when it serenades the ovens at Auschwitz? Or cannot the rocks of a national park be said to lie insofar as their geochemical form has almost imperceptibly been altered by airborne pollution? Lefebvre, by insisting that space lies, insists simultaneously on retaining a poetics of space within a deeply analytical argument.[20] This is a risky move, but I think we have to find a way of doing the same with nature.

Space, Power, and Nature: A Missing Geography

Space is synonymous with power. It has always seemed strange to me, given the long and wide history and geography of this idea— the Greeks and the Aztecs, the conquistadors and Queen Victoria, the Nazis and Alfred Mahan, geopoliticians and political geographers of every political hue— that Foucault is widely cited today as its author (and that this insight is now repeated in such abstract terms). Perhaps no other sign of the spatial turn itself speaks to the widespread ignorance of the history of space and indeed the necessity of such a spatial turn. Lefebvre challenges us to give more than lip service to this collusion of space and power, to mean what we say if we make the connection between space and power and to analyze its consequences.

It was a recognition of the connection between space and power that made the spatial turn so attractive. In its earliest uses in cultural theory the language of space was used to suggest alternative subject "locations" apart from the universal subject; a "territorialization" of experience in alternative domains; a "recentering" of concerns, oppressions, experiences and ways of knowing previously treated as marginal; the "displacement" of the subject; the "mapping" of one experience onto another; the exploration of "situated" knowledges, and so forth. These spatial metaphors adeptly pried open discourses that had previously seemed closed, or simply left them behind, "colonizing" a new space. This rediscovery of the language of space facilitated a politically vital, empowering, and fundamental rewriting of the political with feminism and anti-racist politics especially at the center. It was at its most powerful and most effective when, as with Adrienne Rich's poem, "The Politics of Location," the connection between metaphorical social location and precise spatial location was maintained (she begins the poem with her address as she toyed with it in childhood). And yet very quickly many of these same spatial metaphors have been used up, emptied of meaning, divorced from any space outside the mental (in Lefebvre's terms); the cul-de-sac of identity politics, with everyone in their "own" separate spaces, while not a necessary outcome of such an engulfing idealism, was ushered in by it.[21]

The important task now would seem to be to follow up on the initial recognition of the power of spatial discourse with the harder work of analyzing why space (no longer confined to the mental) is synonymous with power. This, of course, is precisely the terrain of Lefebvre's politics. An outcast and renegade from the official Marxism of the postwar French Communist Party, Lefebvre maintained the importance of class struggle but through his work on everyday life as well as his explorations of the production of space had a long record of what could be called an integral Marxism. He had no illusion that class power had dissipated or been displaced or that the bourgeoisie for its part had given up on class struggle, but he also insisted that through the mantra of "class struggle" the official communist parties had ceased to distinguish them from bourgeois politics as usual. Class struggle remained revolutionary only to the extent that it provoked and promoted "difference," insisted on the "right to difference," he argued, echoing what was becoming a central feminist theme:

... inasmuch as abstract space tends toward homogeneity, towards the elimination of existing differences or peculiarities, a new space cannot be born (produced) unless it accentuates difference. . . . As for the class struggle, its role in the production of space is a cardinal one in that this production is performed solely by classes, fractions of classes and groups representative of classes. Today more than ever, the class struggle is inscribed in space. Indeed it is that struggle alone which prevents abstract space from taking over the whole planet and papering over all differences. Only the class struggle has the capacity to differentiate, to generate differences that are not intrinsic to economic growth *qua* strategy, "logic" or system—that is to say, differences that are neither induced by nor acceptable to that growth. The forms of the class struggle are now more varied than formerly. Naturally they include the political action of minorities (p. 52, 55).

Lefebvre does not solve for us the current burning question of how to put back together in a coherent political vision the fragmented political approaches connected with class and gender, race and sexual preference. The suggested assimilation of multiple forms of struggle to class struggle betrays its own historical roots in the late 1960s. Yet his analysis remains open for such an integration, provides "spaces" for it, while also insisting that class struggle be taken seriously.

And yet the deployment of Lefebvre in this context will have to confront three weaknesses: the awkward hiatus between ontology and history in production of space, the elaboration of what this "right to difference" might mean in practice, and the restoration of a viable politics of nature. In the first place, the conceptual leap between the abstractions of Hegelian ontology and the details of a dramatically changing social and cultural inventiveness around 1910 (to take the most obvious example) is virtually unfathomable; it would seem to take an excessive Hegelianism to implicate the one in the other, but nowhere does Lefebvre explain the connection. The resulting epistemological vacuum invites Einstein's flirtation with ontology, itself occurring in the 1930s, and while Lefebvre is too committed to a socially constructed world to be tempted too far in this direction, the allure of a new "spatialized ontology" has not always been resisted.[22]

This hiatus, however, can in part be filled in by excavating a missing geography, a geography that was effectively buried as it was lived. For Lefebvre is right that a dramatic shift took place around the turn of the century in the social production of space (and time), although it should more properly be seen as spanning several decades from the 1880s to 1918. And it was a more profound shift than Lefebvre's discussion of Picasso would suggest. The extent of this shift is grasped by Kern for whom the changing concepts of space and time in art, technology, communications, and warfare during this period are broadly coincident if not always intimately connected.[23] Where Lefebvre juxtaposes ontology and history, Kern's account juxtaposes different historical conceptualizations; and while the explanatory hiatus is thereby narrowed, it is not dissolved. The sources of historical change in Kern vacillate between the conceptual and the technological. I want to suggest that a vital missing link here may be the equally dramatic shifts that were taking place in the produced geographies of capitalist expansion during this period, a shifting geography that commentators as different as Vladimir Lenin and Cecil Rhodes, revolutionaries and bourgeois geographers, all caught but which, for different reasons, atrophied throughout the twentieth century.

The British geographer Halford Mackinder declared in 1904 that the four century long "Columbian epoch" had come to an end.[24] This was no mere fin de siecle apocalypticism, but tied in closely with the orthodox geographical wisdom that the territorial expansion of European colonialism and hence of capital had ended and that future economic growth would perforce take a different form. Not only Mackinder but his German and U.S. colleagues Alexander Supan and Isaiah Bowman were saying essentially the same thing. If the accumulation of capital at a global scale had previously taken place in large part through the geographical expansion of the sway of capital, this option was now foreclosed. In Lefebvre's terms, this was the period in which abstract space came to fruition at the world scale, taking a particularly Western absolutist form. Hereafter, economic expansion was disconnected from the expansion of the absoluteness of abstract space but was dependent now on the relative expansion of one place or set of places vis-à-vis another. Much as Lenin suggested, economic expansion in the twentieth century was premised on the colonization and recolonization of relative space, the redifferentiation of abstract space from within, as it were, in exactly Lefebvre's sense: the contradictions of abstract space lead to its own redifferentiation. Uneven development proper, as a mode of development of capitalism, only emerges at the end of the nineteenth century.

Now there is a lot to be done with this. It suggests new meanings for the "end of the frontier" discourse in the Americas and South Africa, a new importance to questions of regional development, deindustrialization, underdevelopment, and it casts suburbanization and contemporary urban restructuring—the simultaneous expansion and redifferentiation of urban space—in a very different light. But in a more political vein, it suggests the need to take further Lefebvre's argument about the "right to difference." Assuming that it is not an entirely haphazard process, how is the differentiation of abstract geographical space, by capital itself, organized? And how, in response can a spatialized politics exploit these differences and create new ones that challenge the abstraction of space?

Both within and against abstract space, spatial differentiation occurs in many ways, but in all of this scale is central. Spatial scale is a kind of framework regulating the dimensions of spatial differences that are produced; all space is scale through and within social processes. At its most general, scale represents a spatial resolution—always temporary and shifting—of opposing social processes of competition and cooperation, and is itself socially produced. The production of scale in fact goes to the heart of a politics of spatial differentiation. Scaled space is bounded space and so scale marks the boundaries within which space, quite literally, contains struggle. Yet at the same time, social struggle establishes differentiated spaces at specific scales, and so scale not only becomes the means of bounding struggle but of expressing the ambition of struggle. In the struggle against abstract space, in Lefebvre's terms, what we can think of as "jumping scale" becomes a central strategy: from the body to interpersonal space; from the liberation of a park to the scale of the neighborhood; from the neighborhood to the city; the city to the nation, and so forth.[25]

Finally, there is the politics of nature. The social production of "natural" difference is expressed at every scale from the global to the body and back again, and this is communicated in fragmentary ways in the environmental movement. The differentiation of the global as the scale of environmental "damage" was stressed by the environmental movement that emerged from the 1960s and it has succeeded precisely to the extent that the global ambition of so much mainstream environmentalism has been absorbed by the market. The citizens of North America, Western Europe and Australia now live in a comfortably "environmental" society in which environmental "cleanliness" has become a fetishized consumption good, produced by the massive multinational environment industry (with its various sub-sectors: food, health, recycling, exercise, etc). This fetishism of nature has its appropriate converse in the export of pollution to the "Third World" where, in other spaces, the destructive, fetishistic mimesis of nature also reaches its pinnacle in eco-tourism.

The scaling of environmental politics takes other forms, from the equally fetishized spatial nostalgia of bioregionalism, the proliferation of deliberately single-issue local community organizations, to the ecofeminist insistence on the body as the scale from which a broader environmentalism can be built. But the lack of a sense of the political work done by scale, of the social production of scaled nature, together with the disjointed unity of the different scales of political action themselves, prevents the gathering of these different movements into something more than fragmented tendencies. It prevents the conceptual and practical jumping of scales in pursuit of the most crucial question: what kinds of natures do we want to produce? How can the contradictory differentiations of produced natures be wedged apart in ways that promise the possibility of new, democratic productions of nature—the redifferentiation of an inevitably socialized nature in socially agreed ways?

Lefebvre sought a unitary theory of space for the political purpose of making space anew. More important may be a unitary theory of produced nature that implies such a theory of space, for the purpose of recolonizing nature. And in many ways this might be seen to further Lefebvre's project on its own terms. By the mid 1960s Lefebvre was preparing to return to the studies of everyday life that had occupied his attention since the 1940s, but his momentum was interrupted by the social movements and uprisings of the 1960s, culminating in the "hot spring" of 1968 in which Lefebvre participated. The simultaneity of revolt in different places—Paris, Prague, Newark, Tokyo—and the specifically urban form it took convinced him of the spatial substance of social change, and the desquamation of everyday life could henceforth only be accomplished by an integral decoding of produced space.[26] But in the end it is spatialized nature rather than naturalized space that provides the banal substance of everyday life, and the fruition of Lefebvre's ambition—politically and intellectually—requires an adjustment to the politics of the production of nature.

Notes

1. Henri Lefebvre, *The Survival of Capitalism* (London: Allison and Busby, 1974), 21; originally published as *La survie du capitalisme, la reproduction des rapports de production* (Paris: Anthropos, 1973).

2. For an excellent historical introduction to Lefebvre's political and intellectual life, see the preface by Michael Trebitsch in Henri Lefebvre, *Critique of Everyday Life* (London: Verso, 1991), ix–xxviii.

3. Henri Lefebvre, *Critique of Everyday Life* (London: Verso, 1991) (hereafter *CEL*); originally published as *Critique de la vie quotidienne, I: Introduction* (Paris: Grasset). (The English edition also includes the long foreword from the 2d (1958) ed.) Henri Lefebvre, *The Production of Space* (Oxford and Cambridge, Mass.: Basil Blackwell, 1991); originally published as *La Production de l'espace* (Paris: Anthropos, 1974). Perry Anderson referred to Lefebvre in 1983 as the "oldest living survivor of the Western Marxist tradition" (Perry Anderson, *In the tracks of historical materialism* [London: Verso, 1983], 20. See also Henri Lefebvre, *Writings on Cities*, eds. Eleonore Kofman and Elizabeth Lebas (Oxford and Cambridge, Mass.: Basil Blackwell, 1996).

4. Henri Lefebvre, "Space: social product and use value," in *Critical Sociology: European Perspective*, ed. J. Freiberg (New York: Irvington Publishers, 1979), 291.

5. See Edward Soja, *Postmodern Geographies: the Reassertion of Space in Critical Social Theory* (London: Verso, 1989), 10–75. For a contrary opinion, see John

Agnew, "The Hidden Geographies of Social Science and the Myth of the Geographical Turn," *Society and Space* 13 (1995): 329–80.

6. See Colin Leys, "Underdevelopment in Africa," *New Left Review* (1994); Patrick Bond, *Uneven Zimbabwe* (Harrare: University of Zimbabwe Press, 1996).

7. Casper Weinberger, "Bring back geography!" *Forbes Magazine*, 25 December 1989. See also Lefebvre, "Reflections on the politics of space," *Antipode* 8 (1976): 30–37.

8. Unless otherwise noted, all references hereafter, as noted by page numbers, are to *The Production of Space*.

9. See Henri Lefebvre, *La fin de l'histoire* (Paris: Editions de Minuit, 1970).

10. Or, as Merrifield comments in a very interesting article, "Lefebvre never recanted a certain nihilistic sensibility that he culled from Nietzschian thought." Andy Merrifield "Lefebvre, Anti-Logos and Nietzsche: An Alternative Reading of 'The Production of Space,' " *Antipode*, 27 (1995): 296.

11. Notice that for Lefebvre, contra Foucault, Bergson's era witnesses the demise of one set of spatial codes rather than the origins of a privileged time, *de novo*. For a longer discussion of this transformation—better on time than on space—see Stephen Kern, *The Culture of Time and Space 1800–1918* (London: Weidenfield and Nicholson, 1983).

12. This is precisely the approach that Soja has taken in attempting to reconstruct "spatialized ontologies." See Soja, *Postmodern Geographies*, ch. 5.

13. Lynn Stewart, "Bodies, Visions, and Spatial Politics: a Review Essay," *Environment and Planning D: Society and Space* 13 (1995): 610.

14. See Stewart, "Bodies, Visions, and Spatial Politics: a Review Essay" for an interpetation of Lefebvre as theorist of space through the body.

15. It should be clear that by "the dialectic of nature" I do not mean to resurrect the very vacuous terms in which Engels posed these questions, a form that was further evacuated by official Marxism.

16. See the debate between Benton and Grundmann: Reiner Grundmann, "The Ecological Challenge to Marxism," *New Left Review* 187 (1991): 103–20; Ted Benton, "Ecology, Socialism and the Mastery of Nature: A Reply to Grundmann," *New Left Review* 194 (1992): 55–74.

17. For the fuller elaboration of this argument see, Neil Smith, *Uneven Development: Nature, Capital and the Production of Space* (Oxford and Cambridge, Mass.: Basil Blackwell, 1984), 32–78. See also "The Production of Nature" in *FutureNatural*, eds. George Robertson, Melinda Mash et al. (London: Routledge, 1996), 35–54; Max Jammer, *Concepts of Space* (Cambridge: Harvard University Press, 1969); Margaret FitzSimmons, "In the Matter of Nature," *Antipode* 26 (1989): 12–48; Noel Castree, "The Production of Nature," *Antipode* 21 (1994): 106–20.

18. Karl Marx, *Grundrisse*, trans. Martin Nicolaus (New York: Vintage Books, 1973)).

19. This was the opportunistic slippage at the heart of Alan Sokal's attack on *Social Text*. The "production of nature" thesis in no way implies a suspension of the law of gravity as Sokal assumes, or anything else so silly. That critics could seriously propose that it does only reaffirm the correctness of the social critique

of science which argues that science and scientists are too often locked in narrow and unreflected ways of understanding the world around them (social and natural). See Sokal, "A Physicist Experiments with Cultural Studies," *Lingua Franca* (May/June 1996): 62–64; Stanley Fish, "Professor Sokal's Bad Joke," *New York Times* 21 May 1996.

20. See Merrifield, "Lefebvre, Anti-Logos and Nietzsche," 298.

21. Lefebvre: "Metaphor cannot do duty for thought" (p. 170). For further discussion of this argument see Gerry Pratt, "Spatial Metaphors and Speaking Positions," *Environment and Planning D: Society and Space* 10 (1992): 241–44; Neil Smith and Cindi Katz, "Grounding Metaphor: Towards a Spatialized Politics" in *Place and the Politics of Identity*, ed. Michael Keith and Steve Pile (London: Routledge, 1993), 67–83.

22. A. Giddens, *The Constitution of Society: Outline of the theory of Structuration* (Cambridge: Polity, 1984); Edward Soja, *Postmodern Geographies*, 118–37.

23. Stephen Kern, *The Culture of Time and Space, 1880–1918* (London: Weidenfeld and Nicholson, 1983).

24. Halford Mackinder, "The Geographical Pivot of History," *The Geographical Journal* 23 (1904), 421–37.

25. For this argument in greater depth and in the context of the political struggle for Tompkins Square Park in New York City, see Neil Smith, "Contours of a Spatialized Politics: Homeless Vehicles and the Production of Geographical Scale," *Social Text* 33, (1992): 54–81.

26. cf."The social control of space weighs heavily indeed upon all those consumers who fail to reject the familiarity of everyday life." (*POS*, 233).

The Production of Space *or* The Heterogeneity of Place: A Commentary on Edward Dimendberg and Neil Smith

Edward S. Casey

> Thus space may be said to embrace a multitude of intersections, each with assigned location.
>
> —Henri Lefebvre, *The Production of Space*

I

The production of space? We start witih an oxymoron of major magnitude. For is not space ultimately (or rather, simply and straightforwardly) *given*, and not produced? Given in two ways: first, as the very medium of immediate experience, as the apparently unproblematic "here" from which all "sense-certainty" begins. What could be more self-evident than that we are *in* space, surrounded by it, always already parts of its amplitude and enclosure? How could we produce that in which we stand and move and have our being? Given, second, as a cosmological or epistemological starting point, thus the least likely of anything at all to be produced (since to be produced is to be produced from *something else* already there, from a space before space, engendering a vicious regress).

Such givenness of space is itself a given of Western thought: beginning with Plato's *Timaeus* (where space as *chora* figures as the Necessary, given before Forms are imposed on material qualities by the Demiurge), continuing with Aristotle's *Physics* (wherein space is conceived as *topos*, that is, the container given before, and alongside, what is contained in it), present in Descartes's positing of *res extensa* as the metaphysically given (wholly independent of *res cogitans*), and still evident in Kant's assertion of space as the form of *external* sensibility. These thinkers, and many others in Western philosophy, and counting as well physicists until Einstein, take

71

space not just to be given but given in such a way that the incursions of mind—or any other form of human action, including historical or social interaction—cannot affect it as it is in itself. Even Hegel sees in space (especially the space of Nature) the Spirit utterly outside itself, self-alienated to such an extreme that it cannot act in this most diaphanous of media (for action requires *time*, which is subsequent to space in Hegel's *Philosophy of Nature*).

Beyond this uncompromising prospect of space as given, immune, and pristine—its paradigm found in the idea of void, with which infinite space is often identified in Western reflection from the ancient Atomists to fourteenth century theologians such as Bradwardine and Crescas and on to Newton and Gassendi in the seventeenth century—there is another prospect: space as constructed, made, brought forth. Kant is here the pivotal figure, of course: even if represented as externally situated vis-à-vis the subject, space is still a form of sensibility inherent in that subject. It is *experienced* as given—as "an infinite given magnitude," even as *permanently* given, as the *Critique of Pure Reason* is wont to put it—but it is nonetheless formatively generated from within the same subject, who gives to the given itself its precise geometrical determination. Such constructivism of the a priori is not merely mentalistic, since the geometrical determination is ineluctably historically specific: it is, according to Kant, Euclidean. Thus, even before Hegel attempted to historicize space and time expressly, the Kantian model drives us to history, i.e., to the history of geometry.

But the oxymoron remains: it is one thing to say that the *form* of space is historically instituted (leaving the matter or *Stoff* of spatiality pre- or un-historical), it is another thing to say that space *überhaupt* is produced by human subjects. Thus a first question raised by Henri Lefebvre's book *The Production of Space* is whether he espouses such a radical thesis as that all of space, space of every kind, is produced and (given his own emphasis) socially produced as the title of his book and this symposium implies—and as Neil Smith claims expressly at one point, "space is in any meaningful sense produced in and through human activity and the reproduction of social relations."[1] I would agree with Edward Soja (as cited by Edward Dimendberg) that Lefebvre does not—that for him physical space as such is outside the realm of social production[2] and that his major thesis amounts to the near-tautology that, as he puts it unabashedly himself, "(social) space is a (social) product."[3]

I say "near" tautology because at other moments Lefebvre explores, albeit tentatively, the social determination of natural space, for example in the statement that " 'Nature' itself, as apprehended in social life by the sense organs, has been modified and therefore *in a sense* produced"(*POS*, 68; my italics). This is an equivocal claim in that if the modified element (as what is sheerly apprehended by the sense organs) is merely an epiphenomenon, then its social determination is not impressive but peripheral. The critical question, of course, is the exact meaning of the phrase "in a sense produced"—a sense to which I shall have to return.

II

Before that, let me say how illuminating I found both contributions on which I am here commenting. Each paper complements the other—Edward Dimendberg focusing on the historical and social production of a post- or para-physical space; Neil Smith on such production as it applies to the physical space of nature. Moreover, the two statements treat the two epicenters of Lefebvre's project in *The Production of Space*: History and Nature.

In his incisive essay, Dimendberg focuses on the nature and fate of what Lefebvre calls "abstract space." This is of critical importance, since abstract space is not just another historical avatar of space—despite the fact that it is a member of the sequence of stages of space designated as absolute / historical / abstract / differentiated. It is that member that, as the correlate of Western capitalism, stretches from at least the Renaissance to the present (or, more exactly, to 1910, when things began to change decisively). It is the kind of space within which for the most part we are still entrapped—and all the more entrapped given the endless and tenacious imperialism of global, multinational capitalism. It is thus that with which we must *contend* if we are to imagine a less exploitative experience of space, one that respects the claims of political and social difference.

In the book of mine to which Dimendberg generously refers, I argue that the Greeks (up to Philoponus in the sixth century A.D. as the turning point) endorse a lococentric view of space. Rather than considering space as "absolute"—a term I consider appropriate only in the Neoplatonic aftermath of late Aristotelian philosophy—the ancient Greeks were unusually sensitive to *space as place*, as evidenced in the particularities of given cities and regions (and within these in turn sensitive to the specificities of sacred places, the details of domestic space, etc.). I locate the major revolution in Western conceptions of space just *before* the Middle Ages: this is the idea of *place as space*. From Philoponus onward, and certainly predating the ascension of capitalism, increasing attention was given to the generalities and commonalities of space and, especially, to its absoluteness (i.e., its singularity and irreplaceability) as well as its infinity (its boundlessness, homogeneity, isotropism, etc.). Both traits were exploited in continual efforts—from the anti-Aristotelians active in the Condemnations of 1277 to Spinoza in the late seventeenth century—to identify God

with Space or (as with Descartes) Nature with Space. Early modern thinkers retained the idea of spatial infinity while exploring the relativity of
space (most notably in the case of Leibniz), leading to a preoccupation
with what I like to call "site" in contrast with place, that is, quantitatively
definite, precisely bordered spots of space ideally suited for topographic,
medical, and institutional localizations of the sort examined by Foucault
in *Discipline and Punish* and *The Birth of the Clinic*. On my reading, we
are suffering today more from the pathologies of "site obsession" stemming from the eighteenth century than from the problems of "abstract
space," which on my view reached its apogee in the late seventeenth century. And I believe that a creative response to sited space has begun to
occur in the last half century by the insistent re-recognition of the importance of place by such diverse figures as Bachelard, Heidegger, Deleuze,
Derrida, and Irigaray.[4]

Now I do not retrace this variant history of space and place merely to
differ from Lefebvre but in order to raise the basic question: what if the
history of the subject were less completely, or in any case less exclusively,
correlated with the rise and fall of capitalism—and of forces and social
relations of production—than Lefebvre maintains? What if abstract space
is not only, or not simply, "the tool of domination" (*POS*, 370)? What
then? How would we adjudicate between two such disparate accounts?
On what grounds would we decide as to the validity of the one or the
other? It would be merely circular to support Lefebvre's version on the
ground that space is largely if not entirely determined by social and political factors. But my own account has the problem tha. it reflects mainly
the vicissitudes of *ideas*—i.e., explicit theories of space and place—
without full regard for questions of social and political determination.
The issue is not so much which is the correct account as how shall we do
justice to the ideational *and* the socio-political basis of the production of
space?

This, then, is a first question that arises from a reading of Dimendberg's
provocative essay. A second question concerns the status of the concrete
universal as interpreted by Lefebvre, and it falls into two parts. First, it is
not clear to me why it is the case that, as Lefebvre says (and as Dimendberg cites), "space as actually 'experienced' prohibits the expression of
conflicts" (*POS*, 365). Doesn't the very idea of a *concrete* universal imply
lived contradictions—and thus the real possibility of their expression?
Just as a symptom on Freud's reading is the expression of a lived contradiction in the psychic life of the patient (i.e., a "compromise formation"),
so spatial experience can embody and express social and political contradictions. Surely this follows from Lefebvre's other pronouncement that
"socio-political contradictions are realized spatially" (*POS*, 365). If such
contradictions are realized spatially, then spatial patterns can be consid-

ered as expressions of them, even if not their direct or simple reflections. Second, can we really get away with regarding abstract space as a "degraded" concrete universal—or, alternatively, as an "abstract universal"?[5] If abstract space is a concrete universal of the Hegelian sort (and it is to Hegel that Dimendberg appeals insightfully in his discussion of the *Science of Logic*), then I am not certain that it can ever be strictly *abstract* or altogether *degraded*—given Hegel's view of the dialectical development of the history of Spirit. For Hegel, what is "abstract" is what is comparatively immediate, and thus located at the beginning of any historical process. Hence, if anything, Greek "absolute space" would be abstract on Hegel's criterion. And nothing after this can be entirely degraded if the history of Spirit is genuinely progressive. In other words, there must be *something* about the era of abstract space that realizes what Lefebve calls "the truth of space," however imperfectly, just as there is something about technology that on Heidegger's reading realizes the history of Being in an essential and not merely adventitious way. Even if one does not require the sequence of four stages of spatial history (again: absolute / historical / abstract / differential) to be fully dialectical—a curious lacuna in Lefebvre, as Dimendberg points out discerningly—one cannot leave them as four merely contingent events and still claim inspiration from Hegel, as does Lefebvre. (At stake here, in the end, is the larger question of the history of any important subject matter, whether it be space or time or Being; Heidegger's epochs of Being also look loosely thrown together from a Hegelian point of view.)

III

Where Dimendberg sees a productive source of Lefebvre's thinking in the Hegelian logic of the concept—especially in the generation of the concrete universal from the particular and the individual—Neil Smith decries the "awkward hiatus between ontology and history in the production of space"(ASN, p. 30). By this last phrase, he appears to mean that there is little meaningful communication between the logic of the concept and the actual unfolding of history. I presume that for Smith the production of space is historically orderly, even if it is not ordered by the dialectic of the concept; but this leaves us with the question: by what *other* dialectic does it operate? For it is manifest that Smith does not want to give up dialectics altogether; indeed, his main insistence is that the scope of dialectics include the natural world. Hence his regret that in *The Production of Space* "the dialectic of nature is evacuated" (ASN, 22)—an evacuation that is the direct result, according to Lefebvre, of the hegemony of abstract space. Even if the life of nature is not logical in any Hegelian sense, not even as a concrete universal, it is nevertheless subject to a dialectic that for Smith is resolutely social in character. So much so that we can say that *nature is itself produced*—by the same right as space, and even as a precondition for the social genesis of space. This is Smith's core thesis, one that challenges us to rethink what we mean by "nature" as well as "space."

I welcome this line of thought, as it allows us to draw a critical line around an aspect of Lefebvre's thought that is truly bothersome: an unexamined late Romantic conception of Nature as an original source, a creative force that brings forth unique works and not merely repetitious products. Further, as so conceived, nature exists *for our benefit*: "Nature creates and does not produce; it provides resources for a creative and productive activity on the part of social humanity." (*POS*, 70) Nature "supplies only use value" (ibid.); thus it is held to be so much "raw material," a phrase used throughout *The Production of Space* to denote the ecologically dubious idea that nature is simply there for the taking (cf. inter alia, *POS*, 84, 356.). No wonder, then, that Lefebvre is especially sensitive to the abolition of nature in the regime of abstract space. As Smith puts it lucidly, "he sees only the destruction of nature and not the *production of nature*" (ASN, 24; Smith's italics). Or as Lefebvre says himself, "Nature is left, as it were, in a no-man's land" (*POS*, 376). This is a revealing way to put it, because it shows just how anthropocentric Lefebvre's view of nature is: if the natural world is rendered useless for us, if it is for *no man* (or woman), then it might as well not exist. Despite certain moves that point in a different direction—his description of a peasant cottage as situated between nature and labor, work and product (*POS*, 84)—for the most part Lefebvre holds that "the more a space partakes of nature, the less it enters into the social relations of production" (*POS*, 83). This is an undialectical move indeed, and Smith's effort to bring the dialectic of social relations into the very heart of nature is to be applauded as a courageous corrective to the impasse to which *The Production of Space* brings us.

Smith's courage borders at times on the rash. If Lefebvre goes too far in bending nature to suit human interests, Smith errs in the opposite direction: Nature does not exist in order to *become human*, it is *already human*, already social and historical. Thus Smith claims that natural phenomena can lie, that there are "democratic productions" of nature, and that the independence of natural forces is a "fiction." I find each of these claims questionable: a natural circumstance can be misleading—it can misrepresent an actual state of affairs without for that matter *lying*, which requires the intentional concealment of the truth. The blue sky of Hiroshima on the day of the bombing does not lie; it gives rise to false optimism; a cloud with acid rain misleads us, it doesn't lie; the bird that sings at Belsen does not lie, it sings without knowing the tragic significance of the scene in which it is singing. Nor are the productions of nature "democratic" (ASN, 35); all we can say is that they are put to democratic use. And the forces of nature are often independent of human fate or intention: consider only the movements of tectonic plates, which shift slowly and inexorably and at their own rate—in complete indifference to their effects on human beings.

What is needed is a third position, somewhere between Lefebvre's romanticized and Smith's socialized nature. Strangely enough, each author supplies the crucial clue to this third way: Lefebvre when he claims (in a statement I cited before) that insofar as nature is "apprehended in social life by the sense organs [it] has been modified and therefore in a sense produced"(*POS*, 68) and Smith when he says that "at the scale of everyday life, the form of nature is irreparably produced by social activity" (ASN, 26). That is to say, to the extent that the natural world, which retains its independence of human activity and possesses its own biotic and geological dialectic, enters into human life-worlds (as connoted by the phrases "social life" and "everyday life" just cited), then it is subject to the dialectics of social process. And the same is true of the space that is nature's most inclusive dimension. Just here we retrieve the significance of the phrase left dangling at the beginning in this commentary: space is "in a sense produced"—produced socially—as a phenomenon of the lifeworld of place; it is a feature and function of what I prefer to term the "place-world."

We return thereby to Kant, as Smith may wish to indicate by his reference to "the form of nature." Forms are forms of phenomena. It is as natural *phenomena* that Nature is subject to social production. I do not say "epiphenomena"; there is nothing superficial about the appearances of Nature that are subject to socialization. As Kant himself would insist, the appearances are the very appearances of the things themselves, the phenomena are phenomena of the noumena that are ultimately real—for Kant, metaphysically so; for Lefebvre and Smith, naturally so. Thus the social production of nature goes deep, even if it does not go all the way down through nature itself. And the same may be said for the social production of space.

IV

This brings me to four brief concluding remarks, all of which bear on place—a missing dimension in this discussion.

1. First, the triple distinction between social practice, representations of space, and representational spaces (or, in Dimendberg's more exact

translation "spaces of representations") can be reconceived as a difference between social practices in particular places, cognitive representations of extended spaces (where "space" implies something laid out and continuing indefinitely), and lived places. The lived body is the agent of lived places, and its role in the production of *space*—a role recognized by Lefebvre at numerous points in his remarkable book—is compounded and deepened by its even closer collaboration with the genesis of lived *places*, places whose historicity and sociality are conveyed by the bodies that inhabit or traverse them.

2. Second, place provides a way to meet the challenge to Lefebvre's theory that is so well articulated by Dimendberg: namely, "to analyze abstract space neither as homogeneous (the error of interpreting it as a transparent reflection of the world as immediacy) nor as heterogeneous (the mistake of overlooking how specific contents compose moments of a dialectical unity)" (HFA, 22). Lefebvre himself states that abstract space "*tends toward* homogeneity" (POS, 52; my italics). I would prefer to say that abstract space represents the triumph of space over place in such a way as to favor homogeneity over heterogeneity at every turn, yet not so as to suppress the latter altogether. When Lefebvre speaks of "the polyvalence of social space" (POS, 85), I take him to be referring, at least by implication, to the intrinsic heterogeneity of place. The heterogeneous in space is present as the trace of the historical as well as the prospect of the differential—and both as embodied in particular places whose diversity even in the most hegemonic space remains rebarbative: "home-place as resistance" in bell hooks' sense of this phrase being a case in point. Otherwise put, resistance to abstract space does not come from class struggle alone, even if it is true that such struggle may well "prevent abstract space from taking over the whole planet and papering over all differences" (POS, 55). It is not "only the class struggle [that] has the capacity to differentiate" (ibid.): the same capacity is possessed by place in its individuating and concretizing power.

3. A third remark bears on time and space. The effort to establish primacy between these two dimensions, to rank them as it were, an effort that has characterized modern thought since Kant, shows the insistent need for a return to place as their third term—one that combines both while allowing for the nuances and specificities all too often lacking in the generalities of space and time as general cosmic media. This is so whether we follow Foucault's story of the competition between space and time—nineteenth century temporocentrism giving way to twentieth century spatiocentrism—or Lefebvre's. According to the latter, Hegel first sclerotized time in the space of the state, then Marx, Bergson, and Husserl rebelled by asserting the primacy of time, while in the end (by the middle of the twentieth century) we come full cycle in the era of

abstract space: "the state crushes time by reducing differences to repetitions or circularities . . . Space in its Hegelian form comes back on its own" (*POS*, 23). Lefebvre adds revealingly that at the present moment "society totters while the state *is rotting in place* (*pourrissant sur place*) or reasserts itself in convulsive fashion" (*POS*, 24; my italics). In my view the rediscovery of place in certain twentieth century thinkers is precisely what is capable of resolving the stalemate of the *agon* of space and time. And the place that is rediscovered is not only a scene of rot—such a scene I would again call "site"—but of productive and sensitive differences. Place, in short, ushers in the era of differential space announced by Lefebvre himself. (Hegel, we might note, is once more prophetic: in the *Philosophy of Nature* place figures as the "truth" of space and time.)

4. Finally, it is striking that both Dimendberg and Smith regret Lefebvre's failure to provide a detailed description of what would constitute the "differential space" that constitutes the fourth—and now emerging—stage in the history of space. In Lefebvre's account differential space is present largely by negation and privation: i.e., as born from the contradictions of abstract space, such as quality v. quantity, global v. fragmentary, etc. I would suggest that it is place that can provide the theoretical—as well as practical—basis for a renewed sense of space as differential in character. Does not what Lefebvre calls "the redifferentiation of space" realize itself as place? What else other than place accounts so well for the differences at issue in this fourth kind of space: differences of gender, rights, equality, development (including the "uneven development" on which Smith has written elsewhere)? There is a renewed interest in place in the last part of this century largely because of its capacity to support and exhibit difference so tellingly. And yet Lefebvre—like his two astute readers in this symposium—has precious little to say about place proper: a striking lacuna, yet by the same token a genuine opportunity to open up further dialogue.

But let us leave *le dernier mot* to Lefebvre: "any space [indeed, *any place*] implies, contains, and dissimulates social relationships—and this despite the fact that a space is not a thing but rather a set of relations between things" (*POS*, 83).

Notes

1. Neil Smith, "Antinomies of Space and Nature in Henri Lefebvre's *The Production of Space*," this volume, p. 54. Hereafter designated as "ASN."

2. "Space itself may be primordially given, but the organization, use, and meaning of space is a product of social translation, transformation and

experience." E. W. Soja, "The Socio-Spatial Dialectic," *Annals of the Association of American Geographers*, 70 (1980), p. 210

3. Henri Lefebvre, *The Production of Space*, trans. D. Nicholson-Smith (Oxfordand Cambridge, Mass.: Basil Blackwell, 1991), p. 26; in italics in the text. Hereafter designated as POS.

4. For further development, see *The Fate of Place: A Philosophical History* (Berkeley: University of California Press, 1997), esp. chap. 4–6, 9, 12.

5. Both of these phrases are employed by Dimendberg in his essay, "Henri Lefebvre on Abstract Space," this volume, p. 32. Hereafter designated as "HFA."

Formal Politics, Meta-Space, and the Construction of Civil Life[1]

Mary Ann Tétreault

The division of human activities according to whether they are appropriate to private or public life—to private or public "spheres" or worlds—is a distinction that has been made in Western civilization at least since the classical period.[2] Both gendered and spatial metaphors are reflected in such concepts as "political space" and "public sphere," but we misinterpret the importance of this division of human life space if we imagine that functional roles connected to gender are the sole or even the primary considerations directing the allocation of some portions of it to men and others to women. I argue rather that gender is a marker of a complex system of authority that grants or withholds status and therefore access to the primary arena within which those included compete for especially coveted resources.

However, I argue as well that politics transcends boundaries dividing public from private space and that these boundaries are, themselves, products of politics. Relegating the excluded to a different space puts ideological and institutional barriers in the way of their autonomous participation in the competition for resources of great value, even—perhaps especially—when this competition is defined by the insiders as open and democratic. Citizenship is the primary distinguishing characteristic of legitimate occupants of the public sphere. The expansion and contraction of citizenship rights among the various inhabitants of a state is marked by shifts in boundaries dividing public from private space.

Relationships between status, authority, and gender in the construction of state ideologies defining citizenship are more visible today when we look at "traditional" regimes because rhetorical conventions linking gender to domination and subordination are so often clearly drawn there. Whether, like Morocco's, these ideologies claim ancient roots or, like Kuwait's, they spring from more recent exercises in state-building, the legitimation of dynastic regimes is linked explicitly and simultaneously to female subordination and divine dispensation.³ Under these circumstances, gender is not only a functional category; it is an element in the religious and social tenets and practices that underpin hegemonic ideologies. These ideologies in turn explain and justify existing distributions of political power and the social orders they direct.

> The construction and maintenance of a body of ideological communications is . . . a social process and cannot be explained merely as the formal working out of an internal cultural logic. The development of an overall hegemonic pattern or "design for living" is not so much the victory of a collective cognitive logic or aesthetic impulse as the development of redundancy—the continuous repetition, in diverse instrumental domains, of the same basic propositions regarding the nature of constructed reality.⁴

In this essay I examine the social and symbolic system regarded as the epitome of the democratic ideal, that of classical Athens. Institutionally and philosophically, Athenians created political entitlements and exclusions within their community by segmenting space metaphorically into public and private spheres. Yet despite the prominence of sexual metaphors describing a gender hierarchy dividing autonomous citizens from persons of inferior status, women in Athens were neither universally sequestered nor without power in everyday life. At the same time, the political equality of all male citizens of Athens was far from a social reality.

> The institution of the democracy at Athens brought with it, then, the social production and distribution to the citizens of a new kind of body—a free, autonomous, and inviolable body undifferentiated by distinctions of wealth, class, or status: a democratic body, the site and guarantee of personal and political independence. That, of course, was the ideal; the reality of economic hardship and social dependency was quite different.⁵

The divergence between reality and contemporary as well as historical interpretations that inform our understanding of it, is the focus of this analysis.

The Paradigm: Classical Athens

Among the earliest examples of spatially subdivided virtual realities is the ancient Greek separation of the public space of political life from the private space of the *oikos*, or household. The support of one autonomous citizen required the labor of many individuals: relatives, hired workers, agents, and slaves, and it was this labor that enabled the head of the *oikos* to conduct an autonomous public life.⁶ It provided the material bases for the continuation of individual life—food, clothing, and products that could be sold to earn money—and for the continuation of social life—male children of unimpeachable paternity entitled to inherit their father's place in the community, his obligations to his ancestors, and his material wealth.[7]

Wealthy Athenians displayed their ability to guarantee the paternity of their offspring by sequestering their wives, preventing them from venturing into the public spaces outside the home except for religious ceremonies and funerals.[8] Legally classed with slaves, foreigners, and children, female citizens endured an inferior status. Changes in Athenian law made from 451 to 460 B.C.E. under Perikles, which implied that women as well as men could be citizens,[9] left in place the "social reforms" of Solon restricting the activities of women to the private sphere and granting a woman access to the legal, political, and cultural life of the city only through the mediation of a male *kyrios*, that is, a lord, or "controller."[10] Seclusion did not preserve women from lives dominated by hard physical work.[11] It did prevent them from realizing the income their products commanded in the market, exercising authority over their children—particularly female children who were routinely exposed if their fathers did not wish to spend the money required to rear and then marry them off[12], and controlling property left to them by their fathers.[13]

Athenian homes, the physical location of the private space where women's work took place, are thought to have been dark and squalid domains, a sharp contrast to the elegance and beauty of public buildings and their arrangement in public space.[14] This disparity is the material manifestation of the ideology of separate spheres for men and women. It reflects as well the ideological struggle to direct the allocation of scarce resources embedded in privatization as a state strategy. In a real sense, the home and the city-state competed for the loyalty and support of the citizen. Citizens were required to spend their money on "liturgies," such public works as producing plays, supporting singing groups and athletes, erecting monuments, and equipping warships. They were discouraged from spending money on private consumption, like houses, families, and personal adornment.[15] Liturgies redistributed wealth and minimized competitive private consumption as a way of affirming social status. At the same time, liturgies linked material wealth and political status, providing a strong incentive for the citizen to spend for public works—a kind of potlatch.[16] Citizen sponsorship of liturgies also reinforced the principle of equality. "Demosthenes commended the poverty of private houses in fifth-century Athens and praised the consequent lack of distinction between the homes of its most illustrious citizens and those of the poor."[17]

Directing the economic largess of the citizen away from his family and toward the state also reinforced the obligation of the citizen to serve the state in other ways, chiefly through his own or his sons' military service. K. J. Dover notes in his study of Greek homosexuality that the structure of interstate relations among the Greeks directly affected the nature of familial and communal relationships in a way that promoted the willingness of men to go to war.

> The Greek city-state was continuously confronted with the problem of survival in competition with aggressive neighbors, and for this reason the fighter, the adult male citizen, was the person who mattered. The power to deliberate and take political decisions and the authority to approve or disapprove of social and cultural innovation were strongly vested in the adult male citizens of the community; the inadequacy of women as fighters promoted a general devaluation of the intellectual capacity and emotional stability of women; and the young male was judged by such indication as he afforded of his worth as a potential fighter. . . . [his] strength, speed, endurance, and masculinity . . . were treated as the attributes which made him attractive.[18]

This ethos promoted a highly stylized pederasty among elite Athenians, one consequence of which was its dilution of the father-son bond. A father spent little time with his own family, including his sons, which attenuated his attachment to these sons as individuals. His participation in what Eva Keuls calls an "archetypal homosexual relationship . . . between a childlike or pubescent boy and a mature man. . . . [that] had strong paternal overtones,"[19] provided a substitute limited in time and shaped by normative prescriptions toward an inevitable relinquishment of the son figure to military service.

In contrast to the resources devoted to activities in public space, the home was an inferior object of the citizen's economic expenditure. According to Perikles, "national greatness is more for the advantage of private citizens than any individual well-being coupled with public need."[20] The denizens of the home, on the other side of the public-private boundary, were classed as morally inferior: irrational, uncivilized, and dominated by physical appetites.

> As the *Oresteia* of Aeschylus makes clear, a city-state such as Athens flourished only through the breaking of familial or blood bonds and the subordination of the patriarchal family within the patriarchal state. But women were in conflict with this political principle, for their interests were private and family-related. Thus, drama often shows them acting out of the women's quarters, and concerned with children, husbands, fathers, brothers, and religions deemed more primitive and family-oriented than the Olympian, which was the support of the state.[21]

Such a stark public/private dichotomy served the state in several ways. Denying female citizens access to public life erased the legitimacy of their claims to material resources that could have improved living standards in private space. Women's claims to human resources also were denied legitimacy, whether these were male family members mobilized for warfare or, in the event that politics brought military defeat to Athens, themselves and their daughters as potential slaves, concubines, and corpses. Idealizing the affective bond between an older man and an unrelated young man on the verge of reaching his full potential as a warrior also weakened the claims of families to the bodies of male citizens. At the same time, tying women so firmly to the privatized economy reduced the status value of economic resources spent for private consumption and elevated the status value of spending on liturgies and other public activities.

In some Greek states one or the other aspect of privatization was more limited than in the extreme case of classical Athens. For example, Spartan women oversaw much of the economy but had more political rights than most other Greek female citizens. However, in Sparta, "familial and individual relationships were both formally and effectively subordinated to military organization."[22] The suppression of morally justified claimants to resources coveted by the state was achieved by the nonrecognition of a large meta-space that embraced nearly every aspect of life other than legal marriage and formal politics, even some aspects that, as I shall explain, are remembered as part of the public space only by forgetting their many private qualities.

Meta-Space and Civil Life

Meta-space consists in all areas of human interaction where dimensions of private and public space intersect or overlap, erasing actual though not ideological boundaries that separate life into public and private spheres. Not a purely geographic analogy, "meta-space" also describes a process of contested interpretation to deny a reality that does not conform to the hierarchy of values inscribed by a dominant ideology. In both contemporary and retrospective descriptions of life in classical Athens, such defining elements of Athenian society and culture as the symposium, the market, and religious rites are located in meta-space. There, women as well as men engaged in public activity but the presence of women is now, as it was at the time, elided or forgotten. Ideology masks the heterogeneity of participants and the blurring of what are asserted to be rigid boundaries between public and private that characterize life in these meta-spaces. That life is what I call "civil life," that is, the public life of the private individual.

Along with women's quarters, Athenian houses nearly always included a complex of men's rooms centered on an *andron*, or men's dining room, in which symposia took place. The *andron* was the most luxurious part of the house—for example, it was often the only room with mosaic floors.[23] The *andron* exemplifies the geographic component of meta-space. It was "an intermediate zone between the private domain of the household and the public arena of civic buildings and squares where men spent most of their lives."[24]

Symposia were venues for political repartee as well as for other kinds of entertainment.[25] Resources considered too valuable to waste on the household were lavished on symposia, another indication that these were public events even though they were not formally political. Large amounts were spent on the fine appointments of the *androns*, and to purchase food, wine, and the services of musicians and sexual partners for the guests at symposia.[26] Symposia, with their repartee and speeches, were depicted in Platonic dialogues as arenas where men jockeyed for status and power. This quality is captured by Hannah Arendt in her description of the "space of appearance":

> [A]ction and speech create a space between the participants which can find its proper location almost any time and anywhere. It is the space of appearance in the widest sense of the word, namely, the space where I appear to others as others appear to me, where men exist not merely like other living or inanimate things but make their appearance explicitly. . . . The space of appearance comes into being wherever men are together in the manner of speech and actions, and therefore predates and precedes all formal constitution of the public realm . . .[27]

In Arendt's formulation, the space of appearance is public though at best only proto-political. This ambiguous quality exemplifies meta-space and supports multiple interpretations of its nature and purpose.

The commercial qualities of private space included the virtually cease-less labor of free women and slaves to produce goods and services for the men of the household. Wives were valued according to their capacity for work. Keuls notes the erotic quality of woman at work in the pictorial similarity of the mirror and the spindle on vase paintings, and the sexual fears engendered in men by textile myths such as the stories of Arachne and Penelope.[28] Commerce and manufacturing were part of both private and civil life, some taking place indoors, in the home, and some out-of-doors, in markets. Arendt calls the market "the last public realm," despite its connection to labor and work and its disconnection from politics.[29] Free women as well as free men, along with male and female slaves, worked in public market areas.[30] The wives and daughters of male citizens who did not have enough money to keep slaves to do the outside work shopped in the market and went to public wells to draw water.[31]

Female citizens were not supposed to be involved directly in commerce. However, the linking by the poet Anacreon of elite women engaged in commerce to "breadwomen and voluntary male whores" fails to disguise the fact that Athenian households depended on the labor of their mostly female members, free and slave. It is no wonder that Arendt describes private space as the realm of violence, where the bodies of wives, children, servants, and slaves were used up by the labor that produced the necessities of life for male citizens.[32]

Female foreigners, whether they were married to Athenian citizens or not, routinely worked outside the home in various professions including woolworking, manufacturing, and retail trade.[33] However, the predominant image of women working outside the home, Athenian and foreign, is that they were prostitutes or *hetairai*, "the highest end of the scale of commercially available sexual partners . . . 'companions' or mistresses."[34]

> A . . . distinctive . . . feature . . . of Athenian life in the classical period . . . is the increasingly strict demarcation of the public realm as a male preserve and, thus, as a place of potential exposure and violation for women. Those women who do inhabit the public space are either prostitutes or are assumed to be sexually accessible to men.[35]

However, even though the public world of classical Athens is most generally portrayed as a masculine space, in all but political life very narrowly defined, women and men whose status was juridically impaired because they were foreigners or slaves mingled in the huge meta-space of civil life with citizens of both sexes.

Religious rites also are part of meta-space. Not strictly political, religion and the cultic rites associated with clan membership and civic deities permeated every aspect of life in classical Athens.[36] Religion and the state were deeply interpenetrated, each reinforcing the legitimacy of the other and their joint and several claims on citizens—remember that Socrates was executed following his conviction for impiety toward the gods of the city. Olympian patriarchal religion modeled the exclusion of women from political space.

> [B]y virtue of the homology between political power and sacrificial practice, the place reserved for women perfectly corresponds to the one they occupy—or rather, do not occupy—in the space of the city. Just as women are without the political rights reserved for male citizens, they are kept apart from the altars, meat, and blood.[37]

Yet the incompleteness of the coincidence between religion and politics permitted public participation by women in rites such as the Panathenaic festival and the Eleusinian mysteries, and in special ceremonies to ensure the safety and survival of the state[38], as well as in festivals where women were the primary or only participants, such as the Thesmophoria and the Adonia. In 415 B.C.E., the Adonia was celebrated during a conjunction between preparations for the invasion of Sicily and the publicly visible outcome of the Athenian military defeat of Melos and its impact on the lives of Melian women and children. This coincidence, coupled to the inability (or unwillingness) of contemporary investigations to name a culprit openly, prompts Keuls to suggest that Athenian women, under the cover of the Adonia, were responsible for the mutilation of the herms (phallic statues of the god Hermes), an act that cast a dark shadow over Athenian war preparations.[39] Many find Keuls's interpretation of the mutilation of the herms to be an overly imaginative reading of the evidence available.[40] Even so, a close-to-contemporary play, Aristophanes' *Lysistrata*, encourages us to be skeptical as well of images of Athenian women that deny them a political consciousness and thus the wish, if not the right, to participate autonomously in deciding their own fates.

Sexuality as a Political Metaphor

The interpenetration of public and private is also visible in the regulation of sexual conduct outside of marriage and outside the physical space of the home. Status is a subtext in the heavily sexualized rhetoric of classical Athens and the fundamental determinant of who could participate autonomously in political space. One demonstration of the symbolic connection between sexuality and status is the convention that the *eromenos*, the young male citizen object of a homosexual courtship, was not to permit "penetration of any orifice in his body." By "playing the role of the woman" in a sexual act, "the submissive male rejects his role as a male citizen."[41]

The clearest connection between citizen status and sexuality in classical Athens is revealed in the different treatment of male and female prostitution. All prostitution can be viewed as selling oneself to be used for the amusement or pleasure of another, a surrender of autonomy for money. "It was understood, for example, that a man went to prostitutes partly in order to enjoy sexual pleasures that were thought degrading to the person who provided them and that he could not therefore easily obtain from his wife or boyfriend."[42] A male citizen who became a prostitute, like a lover who would allow his body to be penetrated, also was regarded as rejecting the constitutional protection of his bodily integrity. However, the male citizen prostitute also threatened the integrity of the body politic.

The city as a collective entity was supposedly vulnerable in the person of such a citizen—vulnerable to penetration by foreign influence or to corruption by private enterprise. No person who prostituted himself could be allowed to speak before the people in the public assembly . . . Such a person threatened the coherence of democracy from within and had to be disenfranchised.[43]

A man who violated the bodily integrity of anyone, male or female, citizen or slave, by "hitting, killing, rape . . ." could be prosecuted for *hybris*, the self-indulgent misuse of energy or power.[44] If the victim were a male citizen, "even . . . placing a hand on his body without his consent . . . was an act of *hybris*, or 'outrage'."[45]

Although women, whether citizens or not, theoretically could be victims of *hybris*, female citizens could not be disenfranchised for becoming prostitutes because they had no formal political status. Also, because of their legal subordination to male guardians and their physical seclusion in the *oikos*, their bodies did not enjoy the same degree of protection as a man's. Whether a woman was a prostitute or not, her body was traditionally defined, socially and sexually, as an object of male pleasure. That such phallic pleasure was part of the Athenian male citizen's rights is implied by the establishment under Solon of brothels staffed by slaves where the services of female prostitutes could be cheaply had. "The assumption underlying this . . . reform would seem to have been that a society is not democratic so long as sexual pleasure remains the exclusive perquisite of the well-to-do."[46] The disenfranchisement of male prostitutes and the subsidized supply of female prostitutes reflected the same understanding of democratic citizenship. "[They were] complementary aspects of a single democratizing initiative . . . intended to shore up the masculine dignity of the poorer citizens . . . and to promote a new collective image of the citizen body as masculine and assertive, as master of its pleasures."[47]

The Family Romance of the Division of Space

Lynn Hunt develops Freud's concept of the "family romance" to link individual psychic understandings of power relationships to the ideology of movements intended to alter a particular political regime.[48] Although individuals also produce and consume stories that help them to conceptualize their place in the social order, Hunt uses "family romance" to "refer to the political—that is, the *collective* unconscious . . . images of the familial order that underlie revolutionary politics."[49]

The metaphor of a public sphere of masculine roughness and a private sphere of feminine gentleness and protection is a kind of family romance.

Like other romances, it is virtual rather than real. Yet virtual images are often more vivid, even seductive, than the reality they are supposed to represent. Many prefer the clarity of binary divisions over the ambiguities of a reality in which boundaries are not clear, those one wishes to dominate do not obey cheerfully, and needs as well as desires remain unsatisfied. Wishful thinking induces the substitution of the virtual for the real in people's minds, but what is erased in the metaphor also becomes invisible in the actual. Meta-space is produced by denying a complex reality and substituting for it a simplified family romance. What one "knows" about real actions and their real consequences occupies a meta-consciousness separate from the virtual reality of the family romance. One may know at a deep level that obliterating the Melians is no more rational than Antigone's insistence on burying her brother's body or that a respected Athenian citizen could be giving his baby daughter to slave dealers when he exposes her on a hillside. On the level of consciousness, however, romance rather than reality structures assumptions, perceptions, and, eventually, theories.[50]

Modern Struggles in Meta-Space

The division of life space into two spheres, the public and the private, is more than a strategy for simplifying understanding and analysis. It is a psychological and ideological support for the legitimacy of political and social hierarchies. The family romance of the private space—that it is secure, sheltered, and detached from the potentially soiling activities of public life—dresses up domination as protection. Private space is a black box to which are consigned not only women but all those whose subservience to the household head is the domestic analogue to the subservience of mass publics to rulers and ruling elites. Opening the black box to free its inhabitants requires a new romance, one that tells of the participation in civil life of women and others whose personhood has heretofore been obscured.

In authoritarian systems such as the absolutist states that emerged from feudal northern Europe, only the king was a public person with an autonomous political identity.[51] The destruction of absolutism in a state like France required dethroning the old family romance of the all-powerful father and replacing it with a new romance of fraternal equality.[52] This is a political and a psychological process. Meta-space is the site of contests to control not only entitlement to a public existence, but also "free spaces," actual public settings in which people learn self-respect, assert their identity, and acquire public skills, values of cooperation, and civic virtue.[53]

The battle for entitlement takes place where boundaries between public and private can be shifted to include or exclude specific activities and those engaged in them. Boundaries are moved depending on what the mover wants to be understood as public or private at any given time. Among the interesting examples of this process is the inclusion of Arendt's statement that the market is the last public space in the same work in which she engages in a detailed analysis of the nonpublic quality of labor and work.[54] Indistinctness of boundaries is an inherent quality of meta-space and the struggle to reorganize divisions and the meanings assigned to them. Boundary indistinctness also explains the recurrent contestation in every society that draws lines between private and public space over who belongs where and what the entitlements of each may be. That such battles continue to be fought over people long since dead and spaces long since filled by others living vastly different lives shows that the conflict over autonomy and citizenship conducted through the poetics of spatial and corporeal metaphors continues to have meaning in our own time.

Jeffrey Isaac notes a creative ambiguity in Hannah Arendt's writings about political agency that illustrates this point.[55] Often portrayed as an anti-democrat, Arendt is presented by Isaac as a strong proponent of participatory democracy. For Arendt, this kind of democracy is embodied in decentralized corporate bodies such as labor unions, local governing bodies such as Thomas Jefferson's proposed wards or councils, and in Russian soviets prior to their disempowerment by the Bolsheviks.[56] In contemporary terms this is the democracy of social movements, spontaneous coalescenses of individuals whose interest in specific issues cannot be "managed" through "legitimate political conflict."[57] Isaac argues that Arendt rejects the logic of mass democracy not because it is democratic but because it is oligarchic. She prefers instead a participatory elite politics that, rather than consisting of a small group that rules over the many—that is, a *privileged* group—is a politics that insulates the actors from the many, yet is *unprivileged* by the state.[58] The Arendtian elite is self-selected but it is not anointed. It comes together to empower itself and, in the process, breaks up mass society through voluntary association.

Arendt conceived this idea in her contemplation of the American Revolution and Jefferson's realization "however dimly, that the Revolution, while it had given freedom to the people, had failed to provide a space where this freedom could be exercised."[59] As other critics of liberal democracy also have pointed out, to equate freedom—liberty—with the pursuit of happiness—personal gratification—is to deny civil life.[60] From this perspective, what is commonly defined as "democracy" is only consumption, a solitary choice among mass-produced candidates and platforms by voters "caught up in disempowering, bureaucratic forces largely beyond their control."[61] "Mass democracy" is thus an oxymoron. Neither autonomous nor public, it is difficult even to call the participation of most citizens in mass democracy "politics."

Isaac's insight into Arendt's thought encourages us to reconceive the public sphere of mass democracy as only one aspect of civil life. In the terms I have used here, this framework inscribes the conventional understanding of public space as formal politics whose authoritative character is sanctioned by the state. Center stage is monopolized by the few but, as envisioned by Arendt, this arena has many stages each, at least potentially, an oasis in which every interested person has access to a space of appearance.[62] Two characteristics describe these democratic oases: free space, a physical location where the individual can show her or himself safely; and agency, self-motivated organization to resolve issues in which the individual has an interest in the outcome and can help to create this outcome through speech and action. The oases are part of meta-space whose ambiguous quality allows the activist to move boundaries and shape perceptions to permit action.

To locate citizen activism in meta-space is not to say that those whose access to the public sphere of formal politics is constrained should eschew autonomous participation in the desert of mass politics. However, it is to emphasize that political autonomy and the power to shape public outcomes inhere in human beings wherever they come together. There are myriad places in and from which one can act to satisfy substantive needs as well as fulfill aspirations for a political life.

> [They do] not compete for power. [Their] aim is not to replace the powers that be with power of another kind, but rather under this power—or beside it—to create a structure that respects other laws and in which the voice of the ruling power is heard only as an insignificant echo from a world that is organized in an entirely different way."[63]

The public/private division of life space is far more than a device for engaging in gender oppression. It is a family romance that justifies a particular distribution of power and authority. Women and men can change this distribution by organizing themselves in protected spaces to accomplish directly the goals they want to achieve.

> Freedom, wherever it existed as a tangible reality, has always been spatially limited. . . . The fact that political "élites" have always determined the political destinies of the many and have, in most instances, exerted a domination over them, indicates, on the one hand, the bitter need of the few to protect themselves against the many, or rather to protect the island of freedom they have come to inhabit against the surrounding sea of necessity. . . . [T]his

responsibility touches upon the essence, the very substance of their lives, which is freedom.[64]

The boundary that separates freedom from necessity lies in meta-space. The ambiguity of meta-space does more than disguise the agency of individuals disenfranchised under a particular regime and its state ideology. It also makes possible the construction and protection of free spaces in which those without formal power can come together to empower themselves. In these protected spaces, civil life is conducted and politics is created by autonomous action. Meta-space describes the capacity of people to create spaces of appearance, to come together for the purpose of making themselves explicit in one another's eyes "in the manner of speech and actions."

Notes

1. This essay is adapted from a paper presented at the annual meeting of the International Studies Association, Washington, D.C., 30 March 1994. I thank Valerie Assetto, Roxanne Gile, Antoinette Mercadante, Frank Randall, Dimitris Stevis, Robin Teske, and the anonymous reviewers for *Philosophy and Geography*, particularly the specialist in classics, for their helpful comments on earlier drafts. Research for this work was funded in part by a grant from the United States Institute for Peace. None of those listed is responsible for any errors of fact or interpretation.

2. See, for example, Jean Bethke Elshtain, *Public Man, Private Woman: Women in Social and Political Thought* (Princeton, N.J.: Princeton University Press, 1981).

3. Mary Ann Tétreault, "Gender, Citizenship, and State in the Middle East," paper presented at the Conference on Citizenship and the Middle East, University of Oslo, November 1996.

4. Eric R. Wolf, *Europe and the People Without History* (Berkeley: University of California Press, 1982), 388.

5. David M. Halperin, "The Democratic Body: Prostitution and Citzenship in Classical Athens," in *One Hundred Years of Homosexuality: And Other Essays on Greek Love* (London: Routledge, 1990), 98.

6. Hannah Arendt, *The Human Condition: A Study of the Central Dilemmas Facing Modern Man* (Garden City, N.J.: Doubleday Anchor, 1959), 30–31.

7. Sarah Pomeroy, *Goddesses, Whores, Wives, and Slaves: Women in Classical Antiquity*, hereafter *GWWS*, (New York: Schocken Books, 1975), 60–73, 79–83; K. J. Dover, *Greek Homosexuality* (London: Duckworth, 1978), 88, 149.

8. Pomeroy, GWWS, 80; Alfred Zimmern, *The Greek Commonwealth: Politics and Economics in Fifth Century Athens* (New York: Oxford University Press, 1961), hereafter, GC, 334–38.

9. Raphael Sealey, *Women and the Law in Classical Greece* (Chapel Hill: University of North Carolina Press), 12–13. The author believes that these legal changes were demographically motivated, an issue that Douglas M. MacDowell

believes to be unproven—see *The Law in Classical Athens* (Ithaca, N.Y.: Cornell University Press, 1978), 67–68, hereafter *TLICA*. Whatever their motivation, the new law discouraged marriages between citizens and foreigners, but this came at the price of introducing ambiguity in the citizenship status of men. Both Sealey and MacDowell note that citizenship could be verified by membership in a phratry and a deme, groups that women could not belong to. By requiring that both parents be citizens in order for a child to be a citizen, the new law made it impossible to prove without a doubt that a man was a full citizen—only the male ancestors of his mother's ancestors could be fully certified (Sealey, *Women and the Law*, 14).

10. Halperin, "The Democratic Body," 101; also MacDowell, *TLICA*, 84.

11. Eva C. Keuls, *The Reign of the Phallus: Sexual Politics in Ancient Athens* (Berkeley: University of California Press, 1985), hereafter *TROTP*, 229–31.

12. That child abandonment was a strategy designed to husband household resources is persuasively argued by John Boswell in *The Kindness of Strangers: The Abandonment of Children in Western Europe from Late Antiquity to the Renaissance* (New York: Vintage, 1990).

13. Ibid., 110, 138, 146–47, 232, 260–64; Mark Golden, *Children and Childhood in Classical Athens* (Baltimore: Johns Hopkins University Press, 1990), 87; MacDowell, *TLICA*, 84–98.

14. Susan Walker, "Women and Housing in Classical Greece: The Archaeological Evidence," in *Images of Women in Antiquity*, ed. Averil Cameron and Amélie Kuhrt (London: Croom Helm, 1983), 81–91. Also Pomeroy, *GWWS*, 79; Zimmern, GC, 292; Keuls, *TROTP*, 93–97.

15. MacDowell, *TLICA*, 161–64; Zimmern, *Greek Commonwealth*, 289–90. Zimmern believes that these contributions were voluntary, but MacDowell and also Keuls (*TROTP*, 270) note that providing liturgies was not voluntary and that legal procedures, such as *antidosis*, or the exchange of estates, existed to resolve the claims of citizens who felt that they were required to spend more than others who had greater wealth.

16. Wolf, *Europe and the People Without History*, 186–92.

17. Walker, "Women and Housing in Classical Greece," 82.

18. Dover, *Greek Homosexuality*, 202–03.

19. *TROTP*, 299.

20. Quote by Perikles from Zimmern, *Greek Commonwealth*, 290. Characterization of women by the Greeks from Pomeroy, *GWWS*, 97–103.

21. Pomeroy, *GWWS*, 97.

22. Dover, *Greek Homosexuality*, 202.

23. Walker, "Women and Housing in Classical Greece," 91. The author notes that it is difficult to reconstruct the actual division of space inside Greek houses, but she offers in support of her interpretation evidence from an existing house, located in Kano, Nigeria, where women are sequestered as they were in classical Athens.

24. Keuls, *TROTP*, 163.

25. Ibid., 174–82.

26. Halperin, "The Democratic Body," 110. The meta-space quality of these

expenditures is reflected in the extensive regulation of prostitutes' incomes by city magistrates (*astynomoi*). Keuls notes that Aristotle listed controlling prices for prostitution as the first duty of these officials in his *Constitution of Athens*. Keuls, *TROTP*, 208.

27. Arendt, *The Human Condition*, 177–78.

28. Keuls, *TROTP*, 235.

29. Arendt, *The Human Condition*, 142. Arendt distinguishes between the repetitive and disfiguring tasks of the household, which she calls "labor," and the tasks of the craftsman, which she calls "work." Both labor and work are inferior to action, mutual definition in the space of appearance. The craftsman has artistic autonomy but still is embedded in material reality. In addition, work is private. Thus the craftsman is undefined by action in the public space and therefore is doomed to be forgotten even though what he makes may survive.

30. Yvon Garlan, *Slavery in Ancient Greece*, rev. and enl. edition, trans. Janet Lloyd (Ithaca, N.Y.: Cornell University Press, 1988), 143.

31. Keuls, *TROTP*, 235–40.

32. Arendt, *The Human Condition*, 74–77.

33. Zimmern, *GC*, 338–42.

34. Halperin, "The Democratic Body," 111.

35. Ibid., 102.

36. Louise Bruit Zaidman and Pauline Schmitt Pantel, *Religion in the Ancient Greek City*, trans. Paul Cartledge (Cambridge: Cambridge University Press, 1992), 8–15.

37. Marcel Detienne, "The Violence of Wellborn Ladies: Women in the Thesmophoria," in *The Cuisine of Sacrifice among the Greeks*, trans. Paula Wissingeds, ed. Marcel Detienne and Jean-Hierre Vernant (Chicago: University of Chicago Press, 1989), 131.

38. Joan Breton Connelly, "Parthenon and Parthenoi: A Mythological Interpretation of the Parthenon Frieze," ts, 1994. The standard interpretation of the Parthenon frieze, such as the one that accompanies the display of the Elgin marbles in the British Museum, is that it shows a Panathenaic procession. In her essay, Connelly argues convincingly that the Parthenon frieze depicts the story of the three virgin daughters of Erectheus and Praxithia. Euripedes' play, *Erectheus*, tells how they were sacrificed by their parents to save Athens from conquest by Eumolpos, a son of Poseidon, whose large force of Thracians then threatened the city. (The tendency for interpretations of the meaning of sculptures to conform to the notions of contemporary interpreters is discussed in James Fenton, "On Statues," *New York Review of Books*, 1 February 1996, esp. 36.) Connelly also reminds us of a theme running through the work of Euripedes (see, not only *Erectheus* but also *Iphigeneia at Aulis*) that, just as boys go to war, girls go to sacrifice, both for the good of the *polis*. Equivalent vulnerability to blood sacrifice for the good of the state supports a more active interpretation of female citizenship than is reflected, for example, in the work of Aristotle, and makes the "feminism" of Plato's *Republic* far less of an oddity than it is usually portrayed as being.

39. Keuls, *TROTP*, 23–32.

40. See, for example, Bernard Knox, "Invisible Woman," in *Essays Ancient and Modern* (Baltimore: Johns Hopkins University Press, 1989), 114: "It passes belief . . . that no one noticed it at the time—that organized gangs of phallus-bashing women . . . roamed the streets of Athens in the dead of night without exciting comment. . . . [Keuls's] comment that women were able to circulate by night during the 'counter-cultural' festival of the Adonia will not hold much water: one thing we do know about the festival (and we do not know very much) is that women celebrated it on the roofs of their own houses."

41. Dover, *Greek Homosexuality*, 103–04. Dover notes that homosexual anal violation was commonly used in a wide variety of human societies to demonstrate the subordinate status of newcomers and trespassers, the reality behind the figure of the *herm* with its gigantic erect penis, ready to penetrate a thief of either sex (105).

42. Halperin, "The Democratic Body," 96.

43. Ibid., 97–98.

44. MacDowell, *TLICA*, 129–30.

45. Halperin, "The Democratic Body," 96.

46. Ibid., 100.

47. Ibid., 102–03.

48. Lynn Hunt, *The Family Romance of the French Revolution* (Berkeley: University of California Press, 1992).

49. Ibid., xiii, emphasis mine.

50. The persistence with which the irrational and, by some standards, immoral aspects of classical Athenian culture are ignored by scholars and others who compare modern cultures to it to the latters' disadvantage is a theme of Keuls's book (*TROTP*) and of the 1992 Jefferson Lecture by Bernard Knox (Washington, D.C., May 1992).

51. Patricia Springborg, "Politics, Primordialism, and Orientalism: Marx, Aristotle, and the Myth of the Gemeinschaft," *American Political Science Review* 80 (March 1986): 188; Liah Greenfeld, *Nationalism: Five Roads to Modernity* (Cambridge: Harvard University Press, 1992), 95–97, 109–20.

52. Hunt, *The Family Romance of the French Revolution*. Also, Carol Pateman, "The Fraternal Social Contract," in *Civil Society and the State: New European Perspectives*, ed. John Keane, (London: Verso, 1988), 101–27.

53. Sara M. Evans and Harry C. Boyte, *Free Spaces: The Sources of Democratic Change in America* (Chicago: University of Chicago Press, 1992), ix.

54. Arendt, *The Human Condition*.

55. Ibid., 160.

56. Jeffrey C. Isaac, "Oases in the Desert: Hannah Arendt on Democratic Politics," *American Political Science Review* 88 (March 1994).

57. David S. Meyer, *A Winter of Discontent: The Nuclear Freeze and American Politics* (New York: Praeger, 1990), 1.

58. Isaac, "Arendt on Democratic Politics," 157–58.

59. Hannah Arendt, *On Revolution* (New York: Viking Compass, 1967), 238.

60. This type of criticism is exemplified in Robert N. Bellah, Richard Madsen, William M. Sullivan, Ann Swidler, and Steven M. Tipton, *Habits of the Heart:*

Individualism and Commitment in American Life (New York: Harper Perennial, 1985).

61. Isaac, "Arendt on Democratic Politics," 159. Arendt places consumption in the private space in *The Human Condition*, 109–17.

62. Ibid., 164.

63. Ivan Jirous, quoted in Isaac, "Arendt on Democratic Politics," 163.

64. Arendt, *On Revolution*, 280.

The Stranger on the Green

Luke Wallin

In Richard Sennett's book, *The Conscience of the Eye*, he writes:

> The ancient Greek could use his or her eyes to see the complexities of life. The temples, markets, playing fields, meeting places, walls, public statuary, and paintings of the ancient city represented the culture's values in religion, politics, and family life. It would be difficult to know where in particular to go in modern London or New York to experience, say, remorse. Or were modern architects asked to design spaces that better promote democracy, they would lay down their pens . . . As materials for culture, the stones of the modern city seem badly laid by planners and architects, in that the shopping mall, the parking lot, the apartment house elevator do not suggest in their form the complexities of how people might live. What once were the experiences of places appear now as floating mental operations.[1]

Just as architecture is inscribed with codes for feeling and practice, so are the green spaces at the heart of town, corporate campus, and shopping mall. These "natural"-looking places carry rich emotional associations, as snapshots of both personal memories and a progressive political history. Two expressions from this history, "town common" and "town meeting," enjoy widespread use today, despite the fact that their eighteenth century meanings for the process of democracy-building have been transformed and sometimes squandered. I will sketch the history of some of these transformations and indicate what has been lost and gained. Central to these changes, I suggest, is the evolving concept of the stranger encountered on the public green.

The Democratic Stranger in the Eighteenth Century

According to Sennett, the first uses of the word "public" in English occur in the fifteenth century and identify the public with the common good in society. By the eighteenth century, writers contrasted the public and private domains; for example, Swift used "to go out in publick," to refer to to a society based on this divided geography. In French, *le public* shifted from common good in the Renaissance to a special region of sociability.

In this region one met strangers, persons not defined by known hierarchical codes, but rather who redefined themselves through marketplaces, primarily of capital cities. By 1738, one who moved comfortably among diversity was called in French a cosmopolite. Without inherited wealth or feudal obligation, such a man—not yet a woman—might go forth in a new world to make his way.

In *The Fall of Public Man*, Sennett writes:

> These changes in language were correlated with . . . [those in the] cosmopolis.
> . . . places where strangers might regularly meet grew up. [Such as] . . .
> massive urban parks . . . streets . . . fit for strolling . . . coffeehouses and cafes
> . . . theater and opera houses . . . with the open sale of tickets (instead of
> private control of tickets by the elite).[2]

Of course these developments took place in the *ancien regime*, amidst great social tensions that would eventually erupt into the French Revolution. As Sennett says, this new public life brought society's contradictions to a crisis and created positive opportunities for group life that have yet to be understood.

The excitement that swept both Europe and North America during the late eighteenth century had a great deal to do with cities and strangers; in this matrix the idea of a modern democratic world was formed, with its ongoing dialectic between violent revolution and a new, egalitarian civil life. As people glimpsed new liberties and fraternities, they formed utopian visions of coexistence. They were willing to fight and die overturning old orders, but in the name of peaceful futures with transformed social lives. Space seemed, and was, pregnant with new possibilities.

Proxemic and Dystemic Spaces

In his 1988 book *Space and Spirit in Modern Japan*, Barry Greenbie proposes a pair of descriptive terms for two kinds of public space, which differ according to the welcome offered strangers.

> For the complex array of tribal uses of space that human beings exhibit, I
> have adapted the word *proxemic* coined by the anthropologist Edward T.
> Hall (1966) to designate the study of the human use of space specifically as
> an expression of local culture. To match it, I have invented the term dystemic
> to describe the use of space for more impersonal, abstract relationships that

enable members of various social groups to deal with each other amicably. This word is intended to suggest both greater social distance and great physical scale. *Proxemic* as I use the word is quite close in meaning to the adjective *tribal*, but proxemic relationships can be very modern, as in a professional society, a scientific discipline, a trendy youth culture or a Japanese corporation. They include those values, myths, and rituals that bind together citizens of a modern nation, as well as the more ancient traditions that the word tribal usually signifies. What I mean by *dystemic* is very close to *cosmopolitan*, but all sorts of industrial relationships and places, such as a large factory or an airport, can be dystemic without being what most of us think of as cosmopolitan. . . .

I like to think of dystemic space as the worldwide locale of a community of strangers.[3]

It is important, of course, for anyone to maintain private spaces as well, homes and gardens that feel safe and familiar, free from Otherness and all its threats. Greenbie's book contains an original account of why humans need different kinds of spaces; these he traces to evolutionary history, expressed in the layered structure of *homo sapien* brains. The key point for now is that people require safe habitats that will nurture them emotionally (demanded by the limbic system, or brain level common to all mammals), and yet rich enough to stimulate abstract thought (demanded by the "thinking cap" or distinctively human outer cortex). This means not just the private/public distinction, but that between proxemic and dystemic public spaces as well.

While Greenbie's comparative studies of Japanese and North American places have a strong descriptive base, he does not hesitate to recommend the Japanese treatment of space. In Japan, distinctions among culturally symbolic landscapes are sharp, maintained by principles of design and of customs that correlate behavior with specific sites. In fact, Greenbie celebrates the intense Japanese recognition of the complexities of modern life as correlates of their spatial creations (precisely what Sennett claims the Greeks and Romans had, but modern Westerners have lost).

I want to suggest that it was absolutely essential to those emergent creative possibilities for group life that relatively safe places existed for the interaction with strangers. Markets gathered power and transformed nations, as they do today; but in the late eighteenth century markets were public and diverse, chaotic and relatively easy to enter. The new public spaces of Europe tested the limits of face-to-face learning and peaceful change and gave birth to mass constituencies with their demands for individual rights within bonds of contractual fairness.

One can look at this period as the birthing of modern nationalism, warfare and genocide, but it also stirred the progressive possibilities that are our heritage and our unfinished work. The "market stranger" was, and today often is, the Other—needed just slightly more than feared.

Defining the Otherness of Strangers

Will Wright, in his 1992 book *Wild Knowledge*, calls for broader definitions of key terms by which we attempt to conceptualize the world.[4] These terms include "reason" and "otherness," and his treatment reveals something of fundamental importance about strangers in space.

Wright critiques narrow, "technical" reason, which leads to such medical practices as inserting intrauterine devices (IUDs) and silicone breast implants and performing routine mastectomies. Ordinary common sense is competent to question the health benefits of such ideas and to critique them in the name of what Wright calls "sustainable reason." He rejects the world as described by physics (an imagined realm of objective facts that humans can know) and the account of persons given by economics and psychology (a bundle of desires driven by natural law). These, he argues, are incoherent notions that are destroying the planet.

Sustainable knowledge, on the other hand, exists in a social dimension; it contains no sacred truths, no infallible certainties. And this leads to a new way of conceptualizing otherness, whether embodied in persons or encountered more generally, within language itself. As I put the matter in another essay:

> Wright attempts to spell out the application of these pragmatic ideas via a key concept: the priority of language over individual humans. We can only act as humans through speaking a language. Language, therefore, is logically (not temporally) prior to the existence of any individual human being. A language which is sustainable and ecological will contain the possibility of its own existence, its own continuation. Thus a social self-reference is necessary for collective health.
>
> Finally, Wright moves to the books of Kenneth Burke on rhetoric, for an account of how language works in its socially sustainable dimension. All language is classificatory, positing known vs. unknown terms, the familiar vs. "the other." Otherness is a linguistic necessity, and the tension to resolve it is a necessary condition of living language speakers. Particular classificatory systems will always be somewhat conventional, and none will ever hold a privileged access title to truth and certainty. But some such system will be necessary for talk to continue. Otherness, at its conceptual base, is wildness. The effort by active speakers to articulate and thus tame otherness is a necessary condition of continued language use and social life.[5]

Strangers, appearing in civilized dystemic space, embody the possibility of social changes, of new learning processes. In addition to its *necessary* character, Wright's analysis shows that the stranger (as Otherness, Wildness) is a *relative* concept as well. Human strangers should never be essentialized ("The Arab" or "The Israeli"), nor should linguistic categories ("Oriental" or, for that matter, "Western"). Rather, struggling with the meanings others bring to one's encounters constantly tends to tame the wildness of one's initial impressions. And while individual strangers continuously pass into the realm of the familiar, the context of passage (dystemic space) remains important as the source of new meanings, new wildness.

Finding Dystemic Spaces in America

Today people live and work in clusters identifiable by zip codes that tell advertisers and politicians what they eat and drink, watch and listen to, drive and read. In *The Clustering of America*, Michael J. Weiss claims that the United States contains 250,000 neighborhoods, coded by 36,000 zip codes.[6] This sounds like diversity, but advertisers have learned to organize all these zip codes into just forty neighborhood types. Each has a nickname like "Pools & Patios," "Back-Country Folks," or "Blue-Blood Estates."

Within these zones citizens speak fairly freely to their neighbors, or to people living in "parallel" zip codes: those with similar housing prices, income levels and, most critically, school quality. It is the quest for better schools, with their promise to the next generation, that drives much moving and changing of places. Where is this class segregation broken? Where can one engage in sustained conversations with strangers much different from oneself?

When people cannot meet and talk in safe places, to discover common goals and political strategies, society declines. And it is this loss of possibility, this disappearance of public space in which to engage strangers, that is most troubling in recent transformations of town commons and town meetings in North America.

In the eighteenth century, social change required openness, the air of nature, spaces away from the densely socialized courts and churches of the old order. Whether in New World contact zones or in emergent dystemic urban greenspace, people felt the progressive future gathering on open, common terrain. Today, however, much of nature shrinks before technology and corporate or state control. In Massachusetts, one third of the land is developed or "covered over;" another third is protected greenspace; and the final third is up for grabs. As I discuss below, many town commons have disappeared or been transformed into visual images of their own pasts. And yet one also finds examples of progressive change erupting in surprising new places: the civil rights movement in the "closed" town commons of the south in the 1960s and the community and environmental movements for clean air and water and for sustainable energy in the decades since.

Most surprising in the last decade, as nature has grown more humanized, citizens have escaped into cyberspace. They have constructed a creative political domain, a new dystemic territory filled with outrage and provocation, passion and "flaming" rage, but also with thoughtful discussion much broader than "official" political rhetoric. And list-servs, or discussion groups, on the World Wide Web demonstrate an inspiring power to build coalitions around specific issues.

In this essay I discuss the loss of many traditional common spaces as viable democratic arenas, but I also point to newer alternatives. After consideration of some troubling examples I argue *against* over-generalizing from these; such efforts sometimes stem from nostalgia for pasts that were far from ideal; in extreme or "totalizing" cases, these generalizing efforts sometimes suffer from conceptual confusions between data and interpretation or between first- and second-order language. Recognition of the disappearing political green, I suggest, can lead to strategy as well as despair. And electronic space can reasonably be described as part of Greenbie's "worldwide locale of the community of strangers."

The Stranger as Mugger

Today much of the middle class lives in "edge cities," as Joel Garreau calls them in his book of the same name.[7] In the 1950s, white North Americans began leaving the old downtowns for the suburbs, but men commuted back for work, and women for shopping. In the 60s and 70s, malls moved the shopping out to the burbs, which left only Dad's travels as the family's link to the old urban center. And in the '80s and '90s the jobs also moved. Now the landscape is organized in a new way, with gleaming low-rise corporate buildings out in the cornfields, surrounded by burbs and malls. Many who can afford it migrate from the cities; meanwhile some of old downtowns fester with poverty and crime. A number are truly in ruin, and a stranger there is perceived as a potential mugger, not someone with whom to discuss politics. This separation of city (wasteland) from country (opportunities) is structural in the landscape: highways literally inscribe the division, cutting off inner-city residents from knowledge of, as well as transport to, jobs.

This new reality engenders anxiety about strangers that manifests itself in popular culture. Fear of the freeway offramp as gateway to urban hell has been a premise of a number of recent books and films, such as Tom

Wolfe's novel made into the 1990 Tom Hanks film, *The Bonfire of the Vanities*, Dan Aykroyd's 1991 comedy *Nothing but Trouble*, and Emilio Estevez's 1993 film, *Judgment Night*. Each of these stories begins with elite drivers leaving the highway only to fall prey to the human animals who presumably stalk the bleak streets just beyond, and below, the well-lit freeways that slice through contemporary cities.

Of course, some older downtowns have been restored and remain politically vital. I lived for a long time in Hoboken, New Jersey, which maintains a rich, agonistic street life, in which anything can and will be said (in public parks, public meetings, or anywhere else). Energized by streams of immigrants from regions of Italy, Puerto Rico, and other countries, Hoboken is an old-fashioned city of face-to-face arguments, of hard-fought democracy.

Particular circumstances, however, such as its sources of fresh citizens from abroad, and its proximity to the wealth of Manhattan, make Hoboken atypical. Its second and third U.S.-born generations tend to move out to the suburbs and edge cities, contributing to the trends documented by Joel Garreau and Alexander Wilson (see below). Once these descendants have "made it" in comfortable neighborhoods, they face the same class isolation as other affluent groups.

But what *about* the relatively elite denizens of edge cities—if they are separated as never before from the poor, where do they meet each other, and what kinds of interactions take place there? They seldom meet on the common green, in the old sense of a place vibrant with political possibility, because in many cases it has been destroyed. Its *image* still beckons, but often from across a busy street or from the window of one's own passing car.

Streets and automobiles have frequently severed community life from green spaces; these may continue to exist as conservation lands and viewscapes, but not as political starting points. Nor can people interact on sidewalks in the old sense: the first streets fit for strolling came into being in eighteenth century cities, along with the first parks. But few public sidewalks exist in edge cities.

People find one another in shopping malls, which are often given nostalgic names like Bridgewater Commons and Dartmouth Towne Center. Here, since the malls are private property, political speech is strictly controlled by armed guards; people meet not as citizens in a democratic society, but rather as consumers. Anyone attempting religious or political speech, or even an auditory level above the conversational, can be quickly removed.

These malls, even the very large ones, are extremely safe, and polls show this is what customers want. Here the stranger is only you yourself in different clothing, with the same credit card you carry; she or he does not, cannot, bring any new political possibilities for an open discussion. The stranger, so powerful in eighteenth century Europe, is nowadays oddly under control.

Nevertheless, associations of interactive public space from past centuries gently and insensibly overlay the new North American reality. Mall developers and corporate campus designers assimilate the vocabulary of town common and greenspace, while H. Ross Perot and Ted Koppel and President Bill Clinton speak of their TV shows as National Town Meetings. They evoke in the passive, viewing consumer a sense of good old democratic times from a lost age, as she or he watches in the privacy, and privation, of a living room, a "room for living," away from the public world.

This passive television realm (as opposed to the active world of the web and electronic mail) is an intensification of the role newspapers and novels played in the formation of modern nationalism, as described by Benedict Anderson in his book *Imagined Communities.*[8] When people were separated into their own houses, to sit alone reading both the papers and novels in their local vernacular language, featuring people "just like them," the modern "imagined community" began to exist. Now television offers the private family its ethnocentric repetitions, its vernacular comforts, raising the hot emotions of solidarity with some imagined strangers and of hatred toward others.

The Overlay of Private and Public Images Today

Mention of a town commons puts North Americans in a good mood, one of nostalgia for simpler times when neighbor met neighbor on a safe green in the center of a unified village. Here—goes the myth—common problems were discussed, and each citizen had his, and later her, say. Such images combine associations of nature, designed landscape, safety, and harmony between private and public interests and spaces.

These images have roots in the European peasant communities that eventually became towns, then cities, and that provided the alternative power base to king and pope. The free physical association of farmers and traders—strangers who turned out to be commoners on common ground—gave birth to the ideas of liberty, fraternity, and equality.

J. B. Jackson describes the evolution of some of these emotional associations:

> The lawn was different. It was not only part and parcel of a pastoral economy, it was also part of the farmer's leisure. It was the place for sociability and play; and that is why it was and still is looked upon with affection.[9]

The following passage enumerates ways in which the lawn took both private and public character:

> Maypole and Morris dances never got a foothold in northern America, and for that we can thank the Puritans. But baseball, like cricket in England, originated on the green. Before cricket the national sport was archery, likewise a product of the common. Rugby, and its American variation football, are both products of the same pastoral landscape, and golf is the product of the very special pastoral landscape of lowland Scotland. . . . the English or Irish hunt needs a landscape of open fields and hedgerows.[10]

In the eighteenth century, private garden space formerly controlled by kings moved into the new public domain. At the same time, many social events were bonded in popular imagination with private space; they merged the social—but not the public in a political sense—with the lawn. Jackson writes:

> The lawn, with its vague but nonetheless real social connotations, is precisely that landscape element which every American values most. Unconsciously he identifies it with every group event in his life: childhood games, commencement and graduation with white flannels or cap and gown, wedding receptions, "having company," the high school drill field and the Big Game of the season. Even the cemetery is now landscaped as a lawn to provide an appropriate background for the ultimate social event . . .[11]

Out of this emotional matrix of associations between social, public, and private geographies, a certain contemporary myth was born. North Americans like to imagine a common green where great issues were discussed in an open way: questions like independence from Britain, slavery and national union, women's suffrage and child labor. Indeed, the famous Lincoln-Douglas debates took place on town commons throughout Illinois.

Today many such political town commons have vanished, destroyed by the automobile. In some towns all that remains is a small scenic viewscape, suitable for passengers in their cars to admire without stopping; it is nice for picnics and frisbee, perhaps, but rarely visited by numbers of serious citizens with social questions on their minds. One sees two or three people on a blanket, enclosing an encounter in a temporarily privatized bit of the public space. In Sennett's terms, there is nothing about the park or town center that makes one think of justice, or democracy, or any moral goal at all. The green has become instead a backdrop for an hour's pleasure. You might meet a stranger there for a pick-up game of ball, but not for much more.

The Encroaching Culture of Scene and Spectacle

The scenic commons has joined many other aspects of historical natural and designed landscapes in this shift from interactive to passively viewed terrain. One might consider how "waterview" property has become so valuable that it is mostly privatized and fenced off these days. In Rhode Island, where I live, there are few places where citizens can reach the beautiful beaches without paying fees. At many technically public ocean spots, there is no public parking, a clever means by which privileged residents control magnificent open space.

For another example, consider the five hundred miles of the Blue Ridge Parkway and Skyline Drive in Virginia, constructed during the 1940s so that mobile passengers could enjoy vista after vista from their car windows, and never see a sign of human habitation. This was accomplished by great feats of engineering in building the road, and by removing houses and barns from view. Later, after the "mountaineers" had been educated and cleaned up a bit, some were brought back as tourist attractions; and still later, when this proved unpopular, they were removed again. The old fields cleared by Native Americas and early American settlers were allowed to grow in, to create a "natural"-looking forest for visitors. To drive the park is a thrilling experience, offering endless pinnacles of light and shadow, evergreens and oaks, soaring hawks and tame whitetail deer. Visitors seldom realize they are passing through a vast "nature" constructed specifically for them, refined with close attention to their carefully recorded preferences for the way "unspoilt" nature should appear through speeding glass.[12]

Generalizing the Critique of Visualist Culture

Like Skyline Drive and many beachfronts, town greens often seem examples of spaces that were once richer and more complex and diverse as both nature and culture. Today they are simplified, framed, and policed as apolitical viewscapes, and thus seem part of a gradual transformation of nature into humanized landscapes. This trend is consistent with literary critic Frederick Jameson's critique of modernity, especially his indictment of the commodity fetishism of images of nature. In this section I consider some of his views, and in the following one I examine an alternative, more encouraging way of modeling changes in common space.

Jameson argues that cultural objectification has become the defining feature of the modern world. He speaks of a "historically original consumers' appetite for a world transformed into sheer images of itself and for pseudo-events and 'spectacles.' "[13] Jameson and others have argued that our age, gripped by "late capitalism," is marked by an obsessive rage for "cultural objectification," the creation of spectacles out of the materials of historic styles of architecture, landscape, and cultural practice. This trend toward visual celebration of a status quo (the culture of capitalism) achieves its end through repeated presentation of images that reinforce materialism and individualism, and that, most insidiously, work their magic by their constant presence in our lives as images.

Through the visualist culture of television, movies, magazines, fashion, and period landscapes, we are gently but incessantly kept in a passive, voyeuristic, consumeristic state of mind and emotion. One might say we are suspended in a commodity-fetishist mode of being. Images in sufficient quantity erase narrativity and even thought, and hence eliminate awareness of the historical forces at work producing the detail of our escapist culture. What we are escaping from, of course, is the crisis-ridden capitalism that is producing global wars and environmental degradation along with our local, and temporary, affluence.

This way of thinking and describing cultural spectacles is itself a conceptual model, and as such is subject to critique.

As I put the matter elsewhere:

> Jameson is an admitted determinist, and his rhetorical style signals doom, grieves for the lost cause of a classless society, and celebrates implicitly a "high" culture of intellectuality. His tendency toward totalizing abstraction takes . . . cultural history . . . to a sealed discourse in which ethnicity seems equated with illusion. He joins the ranks of so many analysts of culture who have argued that the most important facts about modern life are those to which ordinary citizens do not have access, whether these are psychological, social or political facts. Such theorists of the human condition often make interesting but apparently unverifiable claims. Nevertheless, Jameson's writing is insightful, and worthy of consideration as an organizing principle for such phenomena as the growing western craving for historic preservation and museum culture.[14]

This includes, of course, the freezing in time of town commons as apolitical viewscapes presented to passing cars.

But now let us turn to an alternative way of generalizing the social and cultural dimensions of common space. As a step in this direction we might consider Habermas's conclusion in *The Structural Transformation of the Public Sphere* that the public sphere has to be "made," it is not "there" anymore.[15] Perhaps this is to say that public spaces grow less dystemic all the time, but that citizens have the power to restore or revitalize them.

Democratic Free Spaces

In *Free Spaces: The Sources of Democratic Change in America*, Sara M. Evans and Harry C. Boyte discuss the modern fate of the eighteenth and nineteenth century ideals of locally based participatory democracy.[16] The model for the older America was the New England town meeting, or the church-centered prairie community meeting, where local politics was both a competent arena for problem solving and a training ground for the skills of civic virtue. By contrast, they worry that modern life is characterized by control over local lives by distant forces—the chain store replaces the corner grocery or bookstore, the urban developer invades a rural town to build condos or resorts, media manipulation erodes voter interest and knowledge of real issues, and so on.

Evans and Boyte are interested in the social and cultural geography of places that continue to keep alive the older paradigm of democratic localism, not through reactive efforts to restore the past, but to build upon that past to create new realities. They cite as positive examples the struggle for women's suffrage, the rise of the Congress of Industrial Organizations (CIO) in the 1930s, the southern civil rights movement of the 1960s, and neighborhood and citizen's movements in the 1970s and 1980s.

They write:

> The central argument of this book is that particular sorts of public places in the community, what we call free spaces, are the environments in which people are able to learn a new self-respect, a deeper and more assertive group identity, public skills, and values of cooperation and civic virtue. . . . These are, in the main, voluntary forms of association with a relatively open and participatory character—many religious organizations, clubs, self-help and mutual aid societies, reform groups, neighborhood, civic and ethnic groups, and a host of other associations grounded in the fabric of community life.[17]

One might add the work done by many historical societies and similiar groups in collecting oral histories and thus preserving the continuity and life of public spaces.

Such ideas are no doubt close to the intuitive radar of many landscape planners and designers, as they attempt to create or shape spaces that will facilitate, rather than elide, genuine citizenly interaction. The question is, are Evans and Boyte simply being optimists, in the face of overwhelming countervailing evidence? And how is such a question to be approached, let alone resolved? Frederick Jameson's overarching gloom, expressed through his historical determinism, could not stand in starker contrast to Evans and Boyte's focus on the power of citizens to create positive change. And this difference does not seem a matter of the data, the first-order language of facts. Both parties might well agree about "what happened" on Skyline Drive, New England beachfronts, and disappearing commons of urbanizing towns. The question is what second-order language of theories with which to organize these facts does one adopt. If determinism and free will are viewed as second-order, or framework, principles, non-decidable by the facts themselves, what critieria for their use are then appropriate?

Determinism is a framework assumption for scientific investigation, but it is a proposition too sweeping to admit of direct testing, evidence, or, least of all, "proof." At the same time, agency, or free will, is a framework assumption of ordinary humans living their lives, especially if they are political activists. To deny agency in this sense is to affirm fatalism, the view that nothing we can do matters. But fatalism is not the same as determinism, nor does determinism imply fatalism. Rather, they are assumptions of different kinds of frameworks or models, and their usefulness (or harmfulness) lies in different contexts, and at different levels of generality. There is no inherent reason why one could not operate deterministically in, say, the context of designing a scientific experiment, and also perform practical tasks as if one's choices and actions mattered, i.e., under the heuristic assumption of agency. Conceptual frameworks (second-order language) can be evaluated by third-order principles, criteria like elegance, simplicity, and long-run sustainability, not by direct conformity to perceptual data.

Conclusions

Where does this leave us with respect to town commons? Earlier in this essay I discussed the trend from emergent eighteenth century common spaces, with their radical democratic possibilities, toward the transformation of these into visual objects denuded of political life. Our foray into alternative efforts to generalize theories of social and political space suggests that such a trend is neither linear nor irrevocable.

The negative second-order view could be stated this way:

In the culture of automobile and television, is it any wonder that actual town commons have lost their moral and political significance but retained their private emotional connotations? One finds, in every direction, the words and images of the historical commons appropriated for consumer spaces. A truncated democracy retreats to the domain of the TV set, the telephone, and the computer, for safe communication between relative elites.

And the stranger? Let him or her come and speak to us on television, where we can remotely control the sound or picture as we wish. Let the stranger become the image of a stranger, exotic and alluring, educational and entertaining, above all powerless to shoot me. Let there be flickering blue light, framing the momentary image of a stranger, evanescent and diaphanous, existing on the edge of silence, beneath the power of my thumb.

But an equally coherent second-order interpretation might run:
While the geography of some town commons seems to slide beneath state and corporate control, passing from vernacular to official landscapes, citizens have proven resourceful in many efforts to forge new democratic (and contested) dystemic zones. These include physical spaces claimed by civil rights and environmental protection movements, but also restored urban centers and the magnificently growing cyberworld of electronic activism. While the loss of a three-hundred-year-old tradition of vital town commons is evident and disturbing, the game is far from over. Those allegedly lost "creative possibilities for group life," mourned by writers like Sennett and Jameson, continue to emerge in new patterns. Physically discontinuous, partially electronic, these new cultural landscapes reveal, in Greenbie's felicitious phrase, the "worldwide locale of the community of strangers."

Notes

1. Richard Sennett, *The Conscience of the Eye: The Design and Social Life of Cities* (New York: Alfred A. Knopf, Inc.; Toronto: Random House of Canada Limited, 1990), xi.

2. Richard Sennett, *The Fall of Public Man: On the Social Psychology of Capitalism* (New York: Random House, Vintage, 1977), 17.

3. Barry Greenbie, *Space and Spirit in Modern Japan* (New Haven, Conn.: Yale University Press, 1988), 53.

4. Will Wright, *Wild Knowledge: Science, Language, and Social Life in a Fragile Environment* (Minneapolis: University of Minnesota Press, 1992).

5. Luke Wallin, "Science and the Paradox of Harmony," *UnderCurrents* (York University, Toronto). 8 (Fall 1996: 40).

6. Michael J. Weiss, *The Clustering of America* (New York: Harper and Row, 1988), 2.

7. Joel Garreau, *Edge City* (New York: Doubleday Anchor, 1992).

8. Benedict Anderson, *Imagined Communities: Reflections on the Origin and Spread of Nationalism* (New York: Verso, 1983).

9. J. B. Jackson, "After the Forest Came the Pasture," 1950, reprinted in *Humanscapes*, ed. Kaplan and Kaplan, (Ann Arbor, Mich.: Ulrich's Books, Inc., 1982), 349.

10. Jackson, "After the Forest," 350.

11. Jackson, "After the Forest," 350–51.

12. Alexander Wilson, *The Culture of Nature* (Oxford and Cambridge, Mass.: Basil Blackwell, 1992), chap. 2.

13. John L. (Luke) Wallin, "Landscape and Community Identity," *Landscape and Land Use* (University of Massachusetts, Amherst) (1989): 5. This draws on Frederick Jameson, *The Ideologies of Theory* (Minneapolis: University of Minnesota Press, 1984).

14. Wallin, "Landscape and Community Identity," 5.

15. Jürgen Habermas, *The Structural Transformation of the Public Sphere: An Inquiry into a Category of Bourgeois Society*, trans. Thomas Burger, with the assistance of Frederick Lawrence (Cambridge, Mass.: MIT Press, 1989).

16. Sara M. Evans and Harry C. Boyte, *Free Spaces: The Sources of Democratic Change in America* (Chicago, Ill.: University of Chicago Press, 1992).

17. Ibid., 301.

Public and Private, Power and Space[1]

Ted Kilian

The literature on public space has been described as a "literature of loss."[2] From Jane Jacobs's concern for the decline of lively public spaces and Richard Sennett's *The Fall of Public Man*, to the essays on the "disneyfi-cation" and privatization of public spaces in the 1980s, and Don Mitch-ell's "The End of Public Space," geographers and others have concerned themselves with the loss of public spaces and the decline of public life.[3] But what exactly has been lost? What, if anything, do these very different observers see in common as "public space?" While some observers strive to reclaim lost public spaces, others have argued that "public" and "pri-vate" are useless or even dangerous categories that should be avoided, combined, or replaced.[4] What is at issue in these discussions? How can an empirical project be undertaken without identifying the "object of analy-sis," i.e., public space?

Geographers have become interested in these debates as part of the broader discussion of the social production of space. I attempt in this paper to advance this discussion in several ways. First, I identify two main strands in the literature on public space and the public sphere. Public space in this literature is emphasized either as a site for impersonal contact or as a site for representation. These approaches are often posed on oppo-site sides of debates over public space, although they share similar con-cerns. I suggest, however, that both approaches are inadequate to explain struggles over public space because they tend, albeit in different ways, to reify their object of analysis (public space) while failing to define it clearly. I argue subsequently that while spaces cannot be categorized as inherently "public" or "private," we cannot and should not collapse or eliminate the concepts of publicity and privacy. It is also insufficient to consider public and private as situated at opposite ends of a continuum. Furthermore, arguing that public space is "produced" still does not avoid the necessity of explaining how publicity and privacy operate within existing, material spaces. I will show that publicity and privacy are not

115

characteristics of space. Rather, they are expressions of power relationships in space and, hence, both exist in every space. Finally I give examples showing how this conception can be used to avoid a problem typical of empirical work in "public space" that almost always begins with a space that is assumed to be public or private, rather than analyzing spaces as sites of both publicity and privacy.

Public Space as a Site of Contact

The literature describing the loss of public space can be traced to Jane Jacobs's concern in the 1960s for the way that poorly planned spaces were destroying a public life she remembered in cities. Public space for Jacobs is a site of contact, but of a certain sort that is neither intimate nor anonymous. She warns against a type of contact she calls "togetherness," which is an ideal that "if anything is shared among people, much should be shared," adding that "the requirement that much shall be shared drives city people apart."[5] Richard Sennett, who shares Jacobs's concern, refers to "togetherness" as "the tyranny of intimacy."[6] In Jacobs's view, three qualities make a city neighborhood "successful": A clear demarcation between public and private space, "Eyes upon the street, eyes belonging to those we might call the natural proprietors of the street," and "Fairly continuous activity."[7] Such informal surveillance and lively activity would make public contact rewarding and safe, encouraging people to engage in public life. Jacobs's and Sennett's concern was that the loss of this public contact would mean the death of public life of the city, accompanied by empty, boring, and dangerous streets.

These ideas continue to hold great currency as numerous writers have applied them to designing or changing spaces to encourage active public life.[8] Her comments were directed mostly at planners with the intention that they would note the inadequacy of spaces designed without the potential for informal public contact. But the response has come mostly (and inadequately) in an "if you build it, they will come" approach to public space, with planners seemingly continuing to believe that certain forms lead inevitably to certain social outcomes, ignoring the ways in which public life forms space.[9]

The success of a project to create such spaces as Jacobs envisioned would derive from contact that is neither intimate nor anonymous. An individual is in the least contact when isolated or anonymous and in the greatest contact among close friends and family.[10] If a public space provides only anonymous or intimate contact, it will fail to generate the informal and impersonal contact that encourages public life. In other words, the subject in this model of public space requires contact on his or her own terms—to meet in public without the requirement for commitment on any other level. "Good" public spaces are then those that assemble rather than disperse, integrate rather than separate, invite rather than repel.[11] As Sennett argues, "The city must be a place where people can learn to join with other people without compulsion to know them as persons."[12] Without such intermediary public spaces, Gehl and others argue, people will retreat into private spaces making only the most necessary forays into the anonymous space of the public, but neither space can provide opportunities for connection beyond an intimate clique of family and friends. One's "community" is limited to the intimate private setting, precluding any connection with those who may be outside that group.

Public Space as the Site of Representation

New approaches to social theory in geography beginning in the late 1980s stimulated a new critique of public space.[13] Greatly influenced by Henri Lefebvre's *The Production of Space*, geographers began to look beyond space as "container for social action" to consider space as a product of "spatial practice" as well as a part of the broader process of production and reproduction.[14] At the same time, publicly owned spaces were being increasingly privatized, meaning both that they were being sold to private owners and that their forms and meanings were being shaped by private interests. In the process of privatization, "undesirables" such as homosexuals and the homeless were being excluded from public space. Those who examined public space in this context were concerned less with contact for its own sake than with representation.[15] Public spaces are seen from this perspective as sites of struggle. Public space is defined not by its use for the public but by the process of its definition, "public space does not designate an empirically identifiable terrain or even space produced by social relations . . . nor . . . concrete institutional sites where meanings are manufactured or circulated. It designates instead the relations structuring the vision and discourse themselves."[16] Although I think that Deutsch overstates the point in denying the materiality of public space, her emphasis on the discursive process of producing space is a valuable addition. What is at stake then, and what must be examined is not only the spaces, nor even the representation constructed within those spaces, but the power relationships that exist within those spaces defining them as public or private and defining users as part of "the public" or part of the "undesirables." As Hannah Pitkin points out, "Our ways of distinguishing public and private, then, are heterogeneous and the question of who gets to do the defining is itself part of the problem."[17] For example, Don Mitchell's "The End of Public Space" focused on the way that homeless people are defined out of "the public." He argued that what they needed was "spaces for representation."[18] Rather than changing the design of public spaces, a change was needed in the processes by which public spaces, and the public that occupied them, were defined. Mitchell describes how the homeless people who use People's Park in Berkeley are not considered "representative of the community." In an effort to exclude these "undesirables," the city and the University of California began a plan to reclaim the park as an orderly, safe (defensible?) space with volleyball courts for college students.[19] What excluded groups like homeless people need, according to Mitchell, is "spaces for representation."

Thus the important issue for "representation" is not so much spaces as they exist but the way that "public spaces" are socially constructed. "Who occupies public space is often decided by negotiations over physical security, cultural identity, and social and geographical community."[20] The reverse is also true. Cultural identity, including who is safe and who is part of one's community, are also deeply affected by who appears in public space. Those who write about the representative role of public space are concerned with exclusions, which are often created by the definition of the "appropriate use" of public space.[21] The importance of public space for those concerned with representation is not contact in a limited liberal sense, but being an active part of "the public" in a political space. The individual alone is un-represented, hence without political power. In Hannah Arendt's terms, such a person does not have the "right to have rights" and does not exist in the "world of appearances," the political space of the public.[22] When an individual or a group is excluded from public space, their needs can be ignored.

When certain actions and spaces are defined as private, the groups associated with those spaces and actions may be marginalized by being categorized as "private." Lynn Staeheli focuses particularly on "women's issues" that typically have been considered private and therefore inferior.[23] The issue for Staeheli is in explaining the mutual and shifting constructions of public and private and how these constructions relate to power. She gives numerous examples of public protest demonstrations where actions that are deemed private, such as breast-feeding and kissing among homosexuals, are enacted in public. This transgression of spatial definitions "pushes back the boundaries between publicity and privacy and the ideologies that construct privacy as inferior and as off-limits for public view."[24]

Critiques of "Public Space"

The two approaches to public space described above, which are often on opposing sides of debates, reflect two meanings of privacy (and publicity by implication). Privacy can signify both privilege and deprivation. In the liberal view of public and private, privacy is privilege; it is power over the space surrounding oneself. As such it is necessary for individuality, confidentiality, and the maintenance of tolerance and pluralism. From this perspective the importance of public space is the provision of contact in a sphere outside the private. Protection of the privacy of the individual is of utmost importance, and public life must be available on one's own terms—on the basis of one's privacy. In the civic republican tradition, on the other hand, privacy is deprivation. It is the absence of power. Power exists only in public space, the site of politics and power. Without access to the public, one has no access to power. According to Hannah Arendt, who is often identified with this position, public life is an essential element of the human condition—without publicity one cannot be fully human.[25]

The liberal "contact" approach to public space has been criticized for its tendency to focus on a public life that is limited and constrained by a bourgeois sensibility. Jacobs's "eyes on the street," especially as developed in Oscar Newman's influential *Defensible Space*, becomes less the empowering activity of a community and more the repressive surveillance of the panopticon.[26] The goal becomes to fill the streets with "normal" users and thereby eliminate "undesirables." William Whyte's public is similarly positioned in a middle class or wealthy, generally male, perspective. His ideal public spaces are free of panhandlers and often good places for "girl watching."[27] Newman and Whyte's ideas were often used in ways that were quite the opposite of what Jacobs had in mind. "Public spaces" were created that were intentionally unusable or exclusionary to prevent them from becoming habitations for "undesirables." Disneyland and New York's Business Improvement Districts are perhaps the ideal defensible spaces.[28] These spaces take full advantage of the sense of loss of public life to create a nostalgic, idealized, and active but also safe and sanitized public space—a "consumable vision of civility" accessible to all who can meet the price of admission and agree to adhere to a strict set of rules of conduct.[29]

This liberal vision of public space reifies both "the public" and public space, masking the internal contradictions that are part of the production of "public spaces." An example of the internal contradiction of definitions of spaces as public or private may be found in Lyn Lofland's definition of public spaces as those spaces "to which, in the main, all persons have *legal access* . . . Public space may be distinguished from private space in that access to the latter may be *legally restricted*."[30] Later in the same work though, this definition is made moot by the description of laws that restrict certain activities in public such as loitering, creating a nuisance, and begging. Such laws have "usefully vague wordings (that) can be used and are enforced quite selectively. In many instances they probably only legalize the practice of spatial segregation that developed independently of the law and that subsequently came to be seen as proper."[31] Reliance on legal definitions is impossibly fraught with contradictions and tautologies as spaces become "public" based on the legalization of exclusions that preceded the law. Mitchell discusses a similar process by which labor protests in the United States were limited when they might potentially disrupt the existing order.[32] In this way, protests were made legal only when they were by definition ineffective.

The "representation" approach began in many ways as a response to these contradictions in the literature on public space. I find this approach much more useful in terms of understanding the actual processes by which people produce and use space in everyday political practice. But despite numerous attempts to operationalize the conception of space as process, when empirical research is undertaken, the "representation" approach often reifies a weakly defined "public space" and suffers many of the same problems as the "contact" approach. Mitchell's description of People's Park, for example, offers two competing images of what public space can be. The normative vision that Mitchell prefers is a space of unmediated political interaction.[33] This is in opposition to the regulated and exclusionary public space envisioned by the (legal) owners of People's Park. Mitchell's claim is that in an anarchic public space, homeless people and other "political movements can stake out a space that allows them to be *seen*. By claiming space in public, by creating public spaces, social groups themselves become public. *Only* in public spaces can the homeless, for example, represent themselves as a legitimate part of 'the public'."[34]

While representation is the goal, Mitchell recognizes that being seen is not the same as being represented as part of a legitimate public. He argues that while the homeless are "all too visible" they are "rarely counted as part of the public."[35] Their publicity is not the same as representation. In reviewing the "reclamation" of Greenwich Village's Jackson Park from "undesirables," Rosalyn Deutsch similarly points out how contestation over public space by homeless people actually enhanced the representation of the homeless as outside "the (legitimate) public . . . seeming to acknowledge public space as conflictual yet disavowing the social conflicts that *produce* space."[36] To the extent that public space is constructed within power relationships that favor powerful institutions of capital and the state, marginal groups are already by definition at a disadvantage in being considered part of "the public." In Lefebvre's terms, the conditions of the "spatial practice" that is created within capitalism are uneven.

Staeheli also finds it difficult to deal adequately with the contradictions of power in space, as she acknowledges when she observes, "understanding that places are constructed through social processes may not provide much analytical leverage when places appear to be constructed as *both* public and private."[37] Staeheli notes that private spaces may be necessary for the protection of marginal groups when, as in Eastern and Central Europe, the power of the state prevented political action in public space. But the groups that organized in private in pre-1989 Eastern and Central Europe did not gain power until they asserted themselves into public space. If the goal then is movement from the powerless private to the public, what happens when such movement yields visibility and vulnerability rather than representation? An exposed, marginalized group is no more represented than it was without visibility. Entry into "the public sphere" indeed may "set the stage for backlash, in which acts intended to outrage are portrayed as typical (of marginal groups)."[38]

When Mitchell argues for a different sort of public space in which no such controls and exclusions occur, he is arguing for a different spatial practice. He envisions an anarchical public space. This is in opposition to the controlled, "disneyfied," privatized public spaces of the volleyball courts and to the elimination of undesirables. Mitchell's normative vision of public space is "politicized at its very core . . . and . . . tolerates the risks of disorder (including recidivist political movements) as central to its functioning."[39] But how can an "unmediated" space be anything but a Hobbesian condition where the most powerful and violent rule? Under conditions of uneven capitalist spatial practice, the homeless would be at a clear disadvantage in this struggle unless it involves a *different* and, as yet, unidentified spatial practice. Mitchell implies an alternative in which space is regulated through a process that is produced internally and "remade by political actors," not unregulated or unmediated (arguably an impossibility).[40] This vision is not unlike a spatialized version of Habermas's public sphere, which is defined and redefined through discursive rationality—political debate.

The public sphere that Habermas envisions as an alternative in which public issues are decided by rational discourse rather than force is also clearly problematic. Habermas has been strongly critiqued for presenting a bourgeois public sphere as "the public sphere," thereby masking the exclusions that it contains. As Jane Mansbridge points out, "the transformation of 'I' into 'we' brought through political deliberation can easily mask subtle forms of control."[41] Similarly, according to Nancy Fraser, the bourgeois public sphere does not only represent an "unrealized utopian ideal" but "also a masculinist ideological notion that functioned to legitimate an emergent form of class rule."[42] Fraser argues that because "the" public sphere has been constructed in an exclusionary manner, the entry into that sphere by subaltern groups implies their acceptance of those exclusionary terms. "Insofar as the bracketing of social inequalities means

proceeding as if they don't exist when they do, this does not foster participatory democracy. On the contrary, such bracketing usually works to
the advantage of dominant groups in society and to the disadvantage of
subordinates."[43] In other words, women and other groups who are not
part of the dominant construction (male, white, bourgeois . . .) must debate in a public sphere that by definition does not recognize their needs.
"Declaring a deliberative arena to be a space where extant status distinctions are bracketed and neutralized does not make it so."[44] What is true
of the public sphere is also true of public spaces. Mere appearance in that
space is not equivalent to *access*: a servant at the master's table is no more
powerful or part of the legitimate "public" in the room than the furniture,
despite the fact that master and servant share the same space. Fraser argues
that by continuing to look toward an ideal public sphere, we look in the
wrong direction for power. Rather than attempt entry into "the" generalized public sphere, we should look toward empowering the existing, competing public spheres that she calls "subaltern counterpublics."[45]

Although I agree with Fraser's assessment of the limitations of "the"
public sphere and the importance of examining other publics, her solution
falls into the same traps as those she criticizes. First, "subaltern counterpublics" are necessarily separate from the spaces of power that make up
"the public." Within her own argument, Fraser admits that these alternative publics must somehow connect within a larger public: "however limited a public may be in its empirical manifestation at any given time, its
members understand themselves as part of a potentially wider public, that
indeterminate, empirically counterfactual body we call 'the public at
large.'" Her argument that "the concept of a counterpublic militates in the
long run against separatism because it assumes a *publicist* orientation and
that "insofar as these arenas are *publics* they are by definition not enclaves" does not give these publics access to publicity and power. Indeed,
she notes that these publics are "often involuntarily enclav*ed*."[46] I disagree
with Fraser in her argument that these alternative publics are sufficient in
themselves to be politically viable. A subaltern counterpublic can either
engage with "the public at large," hence facing the power relations
therein, or remain an isolated group, perhaps as a utopian community.

A second critique of counterpublics is that Fraser ignores the potential
for regressive politics within these sub-publics. She seems to assume that
subaltern counterpublics will not suffer from the same ills as the mainstream public. In fact, counterpublics may reproduce the same inequalities internally or simply act as expressions of power relations produced at
a different scale. As Richard Sennett notes:

"We understand that power is a matter of national and international interests,
the play of classes and ethnic groups, the conflict of regions or religions. But

we do not act upon that understanding. Localism and local autonomy are becoming widespread political creeds, as though the experience of power relations will have more human meaning the more intimate the scale—even though the actual structures of power grow ever more into an international system. The result is that the forces of domination or inequity remain unchallenged."[47]

These counterpublics, insofar as they are public, will surely disenfranchise some people, though at a smaller scale, just as the bourgeois public sphere does. For example, within the women's movement that Fraser identifies as a potential counterpublic, conflicts over race and class have arisen as some women's voices were subordinated within that public.

The Trap

We seem to be left with three equally unpalatable options. The first is that public space as the site of endless Hobbesian struggle in which endless restless fighting is the only outcome. Rejecting this option leaves a difficult choice. Either we must eliminate all inequality before we can engage the political action in public space that would produce an egalitarian, Habermasian public sphere—an impossible contradiction—or we must assume that the public sphere is a sham that masks power relationships, leaving no option but withdrawal into enclaves to avoid co-optation.

This set of contradictory outcomes is largely an artifact of the transition from theory to empirical study in public space studies that maintains a distinction of public and private as separate categories (or ends of a continuum). As Staeheli notes, "the process orientation of theories that emphasize the social construction of space and place as public or private breaks down in some senses when analysts attempt to examine concrete settings."[48]

Both Mitchell and Staeheli choose examples that are thought of in the popular imagination, or explicitly in a legal sense, as "public" or "private," and then expose the contradictions within those spaces. This strategy moves the debate forward by showing how space is defined as public or private through political struggle. However, because public and private are still seen as separate categories (or as ends of a continuum, in Staeheli's case), marginalized groups are seen as not having enough publicity while owners are seen as having too much privacy. The exclusions that exist in public spaces are deemed to make those spaces less public. The solution offered then is to increase the publicity of the marginalized group by transgressing the exclusions that keep them out of the public and by asserting their identities into spaces, thereby claiming it as public.

I would not argue, as others have, for the creation of new categories to replace public and private,[49] that would suffer from the same reification of space as any other categorization. Others have suggested collapsing the categories of public and private altogether, but as Staeheli shows, public and private can be used strategically to empower groups. Collapsing these categories, if it were to become more than an academic exercise, would eliminate a powerful conceptual framework that has served many marginal political movements. What I argue is that public and private are very real and very necessary categories but that public and private *spaces* do not exist as such. This is more than saying, as does Rosalyn Deutsch, that "Any space can be transformed into a public or for that matter a private sphere."[50] I am arguing that publicity and privacy are power relationships that play out in space, and that both publicity and privacy exist as part of all spaces. Ignoring this duality (real physical space and socially constructed publicity or privacy) dooms any definition of public space to insufficiency.

Public and Private as Coexisting Power Relationships within Space

What is required is a method of analysis that does not reify spaces as "public" or "private." I would argue that far from being a contradiction even to the normative vision of public space, privacy—the power of exclusion—is a *necessary* part of all space along with publicity—the power of access. It is impossible to envision a space in which people interact without both exclusion and access as part of its social structure. The "analytical leverage" Staeheli seeks lies in finding publicity and privacy in tension within all space. Privacy is the power to exclude while publicity is power to gain access (see fig. 7.1). The workplaces that Staeheli identifies as private appear private because the publicity of workers is overwhelmed in those spaces by the power of privacy of the owner.[51] Disruption of worker organization entails invading the privacy of workers—where privacy means their power to exclude management intervention—as workers attempt to meet, recruit, distribute literature, and engage in similar activities. Organizing successes come when workers manage to create a counterpublic that by definition excludes other publics, such as management, that is large enough to contest the power to exclude that management otherwise has. This same understanding applies to the labor movements that Mitchell describes.[52]

Public and private space are meaningless terms in the absence of social interaction. To be considered "public," streets, squares, and parks must operate under certain rules and exclusions that paradoxically limit their

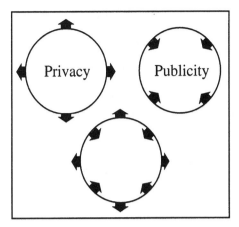

Figure 7.1. Publicity is the power of access. Privacy is the power of exclusion. Both forces operate in tension within the same space.

publicity. People have the right to certain expectations of privacy in public. How can a space be considered public without such restrictions? If a woman "gets what is coming to her" (i.e., is harassed or attacked) for jogging in a park in the dark of early morning, how is that space "public" from her perspective? On the other hand, if all "undesirables" are removed from the park in the name of protecting that woman's rights, the publicity of that park is questionable for those who may be considered "undesirable" on the basis of their race, class, or unconventional appearance. In both cases, public is defined by the idealized (and impossibly contradictory) vision of a space in which privacy—the ability to exclude—is not a necessary component of social relationships. Without the ability to exclude, the ability to limit contact, without boundaries, one is at the mercy of the power of others.

Violations of privacy, whether the privacy of an individual or a group, are violations because they contravene one's right to exert control over some "space." Human rights imply the right to exclusion, at least at the scale of the body. "Privacy is viewed as the means of achieving individuality by providing the barriers necessary to enable the individual to make uncoerced choices in life. Privacy could therefore be viewed as a mechanism for the realization of pluralism and tolerance."[53] The idea that that which is private is privileged does not rely on a liberal individual vision of social relations. Groups as well require such protection for the formation and maintenance of their identities. Counterpublics "function as spaces of withdrawal and regroupment."[54] However, that which is private is also deprived if that privacy is imposed by others.

The power of exclusion and autonomy is necessary to maintain one's rights or one's identity, but these power relationships are maintained through public interaction. Only in a Hobbesian world of "war of all against all" can we imagine a situation in which individuals are entirely responsible for maintaining the boundaries of their private space. Agreements are reproduced in social relationships whether formalized in law or tradition or maintained on the basis of manners, respect, love, fear, or other relations of power. Therefore, unless one has access to the public processes of politics, one's ability to assert privacy is greatly limited.

Counterpublics also function as "bases and training grounds for agitational activities directed toward wider publics."[55] As discussed earlier, if one cannot access the public sphere on one's own terms, then there is no real access. Fraser's counterpublics are "public" in that they are accessible to those who identify themselves in similar ways, but I would argue that they are also necessarily "private" inasmuch as they exclude the discourse of the dominant public sphere (and perhaps other subaltern counterpublics as well) in favor of their own discourse. Without this power of exclusion how can these groupings exist at all? They may be privileged within themselves, and therefore empowering, but so long as they remain outside "the public at large" they are deprived as well, unable to effect social change. Access to political power on one's own terms requires both the power of exclusion (privacy) and the power of access (publicity).

In order to understand publicity and privacy as categories of power, power must be understood as a relationship rather than a commodity. Power is a relationship among social actors, not a commodity that is given or taken away or of which one has more or less.[56] One cannot be in the category "powerful" or "powerless," rather we exist in power relationships with others.[57] This relationship is constantly contested and reproduced so one may have more or less power of exclusion or access in any space depending on the power relationships that exist in that situation.

Hence what Staeheli sees as "pushing back boundaries of public and private" is more accurately a thinning of those boundaries.[58] The limitations of being forcibly relegated to the private are diminished, but so are the protections provided by privacy. What gay protesters at kiss-ins and publicly nursing mothers seek is not an increased publicity but an increased public legitimacy of their privacy. In fact, they already suffer from too much publicity inasmuch as their private affairs are proscribed by the public. Increased publicity only means exposure and vulnerability unless it is based in the power of a privacy that allows a group to maintain its identity. Staeheli states as much when she notes that public nurturing "instilled a sense of public responsibility in maintaining an act as private."[59] Mitchell similarly notes that the homeless people in People's Park are fighting to maintain their identity—their difference, which is threatened with absorption by the state and capital.[60] What the homeless people in Mitchell's example need more of is *both* publicity—through which their needs can be recognized as legitimate—and privacy—through which they can protect themselves from absorption or delegitimation by the public.

This does not mean, as Staeheli suggests, that public and private lie along a continuum. It would be wrong to follow this logic by placing public and private at opposite ends of a power relationship. Those who have the greatest power over space have *both* the greatest power of access and the greatest exclusion.[61] One can have virtually unlimited power of access, but without the power to exclude, that access will be at the mercy of those who define public space. Unless a homeless person has privacy rights, no amount of publicity will give them meaningful access to the public and the right to define the appropriateness of their behavior or their identity in public. In fact, for many homeless people the anonymity of the street, which provides some privacy in isolation, is preferable to life in shelters which, although enclosed, take away privacy through excessive rules and regimentation of contact. Neither option provides any protected access to the public sphere as both marginalize homelessness without acknowledging the "right" of the homeless "to have rights."

Conversely, complete power of exclusion means nothing without the power to access the public sphere. During the years leading up to the Velvet Revolution in 1989 in Czechoslovakia, Vaclav Benda used the term "parallel polis" to describe a protected, alternative public sphere.[62] Similar to Fraser's subaltern counterpublics, the parallel polis was the site of underground opposition. It was public in that it provided access to a public sphere, but also necessarily underground, separate, protected, and therefore private. In the spaces of the parallel polis, power to access "the public at large" was built up within these "private" (protected) organizations. Although they built strength within these enclaves, it was not until they confronted the hegemonic power of the "Communist" regime that they achieved any social change. This was done in incremental form with the release of the "Charter 77" declaration and the publication of *samizdat* press, but the moment of revolution came when the parallel polis appeared on the streets and squares of Czechoslovakia in late 1989. Acknowledging this transformation, one group, which called itself the Movement for Civil Liberties, issued a statement just before the student marches that began with the line "The time has come to get involved in politics."[63]

We do not move from public to private, rather we are constantly within both, simultaneously protecting ourselves from absorption into the public through the power of privacy (exclusion) and asserting ourselves into the public sphere (the realm of political power). Without access to the public, we are politically and socially marginalized, but without a basis in privacy, we enter the public sphere with no basis—we are there without being there.

Toward a New Conceptualization?

The importance of protected but visible space is explained in Lefebvre's argument that social movements must produce their own space. He is arguing for control of the "differential spaces" that are produced in opposition to the hegemony of "abstract space."[64] To have any power or ability to contest domination, social movements must have spaces in which they have both the right of privacy and the right of publicity. This requires the ability to access and challenge the dominant discourse, but in terms of a self-defined and protected alternative. The "differential spaces" in which "abstract space" may be challenged can be different only if they have the power to exclude the dominant representations of space.[65] Differential space must have the protection of privacy. At the same time, the potential for a space to be emancipatory depends on the power of access. So long as a "spatial practice" remains marginal, it can lead only to marginal and easily absorbed or deflected changes. The challenge to dominance may originate on the margins, but it must enter the mainstream (on its own terms) to be effective.

This vision of public space implies a different empirical project than the one that traditionally has been undertaken. It is worthwhile to consider "public" (as well as "private") spaces, but only inasmuch as we consider the power relationships of publicity and privacy at work within all spaces. Studies that decry the loss of public space should recognize that this is not a change simply in the publicity of that space, but a change in the power relationships of publicity and privacy.

The material spaces remain no less important in this conception. In the earlier example of the Velvet Revolution, as well as in Hershkovitz's analysis of the Tiananmen Square student uprising in 1989, material spaces are significant sites of contested meaning that cannot be ignored or replaced by electronic communication.[66] "Ideologies, or as Foucault puts it, discursive practices, are created in specific spaces. These spaces provide the pictures in our minds when we conceive our identities. In turn, ideologies structure, and continue to structure, the ongoing production of spaces; the distinctions between high and low, sacred and profane, gentrified neighborhood or inner city."[67] As the material realization of ideology, material spaces expose the "contradictions in space" that necessarily precede the formation of alternatives.[68] "The built environment materializes meanings—sets them in concrete and stone. In the process of making meaning material, these images become open to question and challenge."[69]

Toward a Methodology

Having argued that public and private are power relations that exist simultaneously in space, and that space is itself socially produced, how can we move toward empirical study that doesn't reify spaces as public or private? How can we examine both material spaces and social processes without absorbing either one into the other. In *Buildings and Power*, Thomas Markus provides a framework for understanding power constellations among three categories of people that buildings serve to "interface." This framework can be used for understanding relationships within any space. For our purposes we can rank Markus's groups in terms of their powers of access and exclusion:

Inhabitants: These are the controllers who have rights to access and exclusion. They may be owners or may be otherwise empowered by the state to exert exclusionary control over space.

Visitors: These are the controlled. They "enter or stay as subjects of the system."[70] They have rights to access for specific "appropriate" purposes, subject to approval by the inhabitants, and have no rights to exclusion.

Strangers: These are the "undesirables." They have no rights to either access or exclusion and are in fact excluded by definition.[71]

These are fluid categories whose members change depending on the space in question. For example, in your home, you may claim the rights of inhabitant with complete rights to access at any time and for any reason. Visitors are only those whom you invite for a purpose and they have no recourse if you choose not to invite them (even if on the basis of race or some other arbitrary measure). Anyone who is not invited by you is a stranger and you have complete power to exclude them. Such spaces are more "private" because the inhabitants have almost complete discretion over visitors and strangers. The decision about who is a visitor and who is a stranger is almost completely at the discretion of the inhabitants. Of course, the state may allow access by others to even this most private space against the inhabitant's will under various conditions such as search warrant or repossession.

A restaurant, on the other hand is more "public" in its power structure. The employees are ranked as inhabitants who may have greater or lesser degrees of access and rights to exclude. The owner or manager may have access to all spaces in the restaurant, but a waiter or busboy may not be allowed in the office. Similarly, a waiter will not likely ask a "stranger" to leave, but rather call a manager. Customers are visitors who are allowed to enter to eat, because that is the appropriate use of the restaurant. Others may come in as visitors too, but if they are outside the "appropriate use" of the space, they may be treated as strangers. For example, if someone comes in looking for a bathroom or to come in out of the rain, she may be allowed to remain as a visitor for that purpose, but the inhabitants (restaurant management) can also deny her entry if they so desire. Strangers are defined by the inhabitants in accordance with laws and norms of the society. The appropriate use of the space is clearly defined, and inhabitants have broad powers of exclusion within that definition; they do not have, however, the nearly absolute rights of exclusion that exist in a private home. The lunch counter sit-ins of the 1960s in the United States were a challenge to the right of inhabitants/owners to exclude visitors on the basis of race. It was decided, by law and in the face of social pressure, that visitors (i.e., restaurant customers) could not be excluded, but strangers (i.e., people who enter as other than customers) could be.

A park is structured as even more "public" than a restaurant. No one group has total power of exclusion except the state. The police and park employees, as agents of the state, have the power to act as inhabitants. Decisions about who is a stranger and who is a visitor are not made arbitrarily by the police, but by a set of laws that define (albeit vaguely) appropriate and inappropriate uses. Because these decisions are made on behalf of inhabitants vaguely defined as "the public," they are open to debate. And because laws often normalize existing exclusions, marginal groups may be deemed by law to be inappropriate. Loitering is difficult to define in a park full of unstructured activity (hence the installation of volleyball courts in People's Park helps define "appropriate use" more exactly).

The difference among these cases, and their resultant level of "privacy or publicity" has to do with the power that inhabitants and visitors have in defining strangers, or inappropriate users. In other words, the definition of powers of exclusion or access depends on how one is categorized within these groups. In the most "private" spaces, inhabitants have almost complete discretion over both access and exclusion, strangers have, by definition, no power of access or exclusion, and visitors have access but virtually no voice in defining the space. The more "public" a space is, the more power its visitors and strangers have to contest their status and the appropriate use of that space.

This approach implies a different political project as well. If we argue for the right of subalterns to "access public space," they enter at the mercy of the "inhabitants"—those actors who have more power over exclusion and access and the power to define the "appropriate use" of spaces.[72] We find ourselves in the unenviable position of arguing for the rights of the homeless to sleep in parks, instead of for their human right to privacy—to a home. If we argue against surveillance in public spaces, how do we answer those who argue that prior to surveillance in a space the women, children, elderly, and other groups were afraid to use those spaces? The focus of this argument should be against the process of surveillance. Who surveys whom, not whether or not there is surveillance. Like the position of being intolerant of intolerant people, a policy of eliminating exclusion can only become a process of replacing one exclusion with another. Rather than seek an ideal public sphere that is free of exclusion, power, and privacy, we should focus on the processes in which the necessary contestation of privacy and publicity is played out. This is not a utopian outcome, there is no end point; in fact, it is arguably anti-utopian inasmuch as it implies an inescapable struggle within power relationships, that may change, but cannot be eliminated without absorbing difference.

Notes

1. My thanks to the members of the Rutgers Geography "Public Space" reading group and to Bob Lake and Don Mitchell for many very helpful comments.

2. Michael Brill, "Transformation, Nostalgia, and Illusion in Public Life and Public Place," in *Public Places and Spaces*, ed. Irwin Altman and Ervin H. Zube (New York: Plenum, 1989), 8.

3. Jane Jacobs, *The Death and Life of Great American Cities*, (New York: Random House, 1961); Richard Sennett, *The Fall of Public Man: On the Social Psychology of Capitalism*, (New York: Random House, Vintage, 1977); Michael Sorkin, ed, *Variations on a Theme Park*, (New York: Hill and Wang, 1992); Sharon Zukin, *The Cultures of Cities*, (Oxford and Cambridge, Mass.: Basil Blackwell, 1995); Don Mitchell, "The End of Public Space," *Annals of the Association of American Geographers* 85, no. 1 (1995): 108-33. I should note that Mitchell's work has been very important to my own understanding of public space. I critique his work here not because I find it lacking, but because I find it to be a fruitful starting point.

4. Liz Bondi, "Geographical Perspectives on Women Specialty Groups: Rethinking Public/Private Space" (paper presented at a panel of that name at the National Conference of the American Association of Geographers, Charlotte, N.C. April 1996).

5. Jacobs, *Death and Life of Great American Cities,* 62.

6. Sennett, *The Fall of Public Man*, 338.

7. Jacobs, *Death and Life of Great American Cities*, 35.

8. Jan Gehl, *Life Between Buildings: Using Public Space* (New York: Van Nostrand Reinhold 1987); Stephen Carr, Leanne G. Rivlin, Mark Francis, and An-

drew M. Stone, *Public Space* (Cambridge: Cambridge University Press, 1992); William H. Whyte, *City: Rediscovering the Center* (New York: Doubleday, 1988).

9. This and many other ideas came out of the reading group on public space at Rutgers University in 1995–1996.

10. Gehl, *Life Between Building*, 17.

11. Gehl, *Life Between Buildings*.

12. Sennett, *The Fall of Public Man*, 339.

13. Edward W. Soja, *Postmodern Geographies: The Reassertion of Space in Critical Social Theory*, (London: Verso, 1989); Derek Gregory, *Geographical Imaginations*, (Oxford and Cambridge, Mass.: Basil Blackwell, 1994).

14. *The Production of Space*, trans. Donald Nicholson-Smith (Oxford and Cambridge, Mass.: Basil Blackwell, 1991) was only translated into English in 1991, so the introduction of Lefebvre's ideas to an English-speaking audience came largely through the work of David Harvey, especially *Social Justice and the City* (London: Edward Arnold, 1973). The quotes are from Lefebvre, *The Production of Space*, 38.

15. See especially Sorkin, ed, *Variations on a Theme Park*; Rosalyn Deutsch, "Art and Public Space: Questions of Democracy," *Social Text* 33 (1992): 34–53; Mitchell, "The End of Public Space?" Contact remains an important part of what I am calling the representation perspective, but it is most significant as it relates to the power of groups to be a part of defining the public and public space.

16. Deutsch, "Art and Public Space," 43–44, italics added.

17. Hanna Fenichel Pitkin, "Justice: On Relating Private and Public," *Political Theory* 9, no. 3 (1981): 329.

18. Mitchell, "The End of Public Space?", 115.

19. Mitchell, "The End of Public Space?", 110.

20. Zukin, *The Cultures of Cities*, 24.

21. Tim Cresswell, *In Place/Out of Place: Geography, Ideology, and Transgression*, (Minneapolis: University of Minnesota Press, 1996).

22. Hannah Arendt, *The Human Condition* (Chicago: University of Chicago Press, 1958).

23. Lynn Staeheli, "Publicity, Privacy and Women's Political Action," *Environment and Planning D: Society and Space* 14 (1996): 601–19.

24. Ibid., 610–11.

25. Arendt, *The Human Condition*.

26. Oscar Newman, *Defensible Space; Crime Prevention through Urban Design* (New York: Macmillan, 1972).

27. Whyte, *City: Rediscovering the Center*, 28.

28. See especially Sorkin, ed, *Variations on a Theme Park*, Mike Davis, *City of Quartz* (New York: Vintage, 1992), and Zukin, *The Cultures of Cities*.

29. Zukin, *The Cultures of Cities*, 26.

30. Lyn H. Lofland, *A World of Strangers: Order and Action in Urban Public Space* (New York: Basic Books, 1973), 15–19; italics in the original.

31. Lofland, *A World of Strangers*, 71.

32. Don Mitchell, "Political Violence, Order, and the Legal Construction of Public Space: Power and the Public Forum Doctrine," *Urban Geography* 17, 2 (1996).

33. Mitchell, "The End of Public Space?", 115.

34. Ibid., 115; italics in the original.

35. Ibid., 118.

36. Deutsch, "Art and Public Space," 39; italics in the original.

37. Staeheli, "Publicity, Privacy and Women's Political Action," 607; italics in the original.

38. Ibid., 611.

39. Mitchell, "The End of Public Space?", 115.

40. Ibid., 115.

41. Jane Mansbridge, "Feminism and Democracy," *The American Prospect* 1 (Spring 1990): 127.

42. Nancy Fraser, "Rethinking the Public Sphere: A Contribution to Actually Existing Democracy," *Social Text* 25/26 (1990): 116.

43. Ibid., 120.

44. Ibid., 115.

45. Ibid., 123.

46. Ibid., 124.

47. Sennett, *The Fall of Public Man*, 339.

48. Staeheli, "Publicity, Privacy and Women's Political Action," 607.

49. Doreen Mattingly, "Geographical Perspectives on Women Specialty Groups: Rethinking Public/Private Space" (paper presented at a panel of that name at the National Conference of the American Association of Geographers, Charlotte, N.C. April 1996).

50. Deutsch, "Art and Public Space," 39.

51. Staeheli, "Publicity, Privacy and Women's Political Action," 614.

52. Don Mitchell, "Political Violence, Order."

53. Judith Squires, "Private Lives, Secluded Places: Privacy as Political Possibility," *Environment and Planning D: Society and Space* 12 (1994): 390.

54. Fraser, "Rethinking the Public Sphere," 124.

55. Ibid., 124.

56. Robert W. Lake, "Negotiating Local Autonomy," *Political Geography* 13 (1994): 423–42.

57. Michael P. Brown, "The Possibility of Local Autonomy," *Urban Geography* 13 (1993): 257–79.

58. Staeheli, "Publicity, Privacy and Women's Political Action," 610–11.

59. Ibid., 613.

60. Mitchell, "The End of Public Space?", 124.

61. Staeheli notes this on page 607, but it contradicts her conception of public and private as continuum.

62. Vaclav Benda, "The Parallel Polis," in *Civic Freedom in Central Europe*, ed. H. Gordon Skilling and Paul Wilson. (New York: St. Martin's, 1991).

63. Gale Stokes, *The Walls Came Tumbling Down: The Collapse of Communism in Eastern Europe* (New York: Oxford University Press, 1993), 153.

64. Lefebvre, *The Production of Space*, 370–72.

65. Ibid., 373.

66. Linda Hershkovitz, "Tianenmen Square and the Politics of Place," *Political Geography* 12 (1993): 395–420.

67. Zukin, *The Cultures of Cities*, 293.

68. Lefebvre, *The Production of Space*, 292 ff.

69. Cresswell, *In Place/Out of Place*, 47.

70. Thomas A. Markus, *Buildings and Power: Freedom and Control in the Origin of Modern Building Types* (London: Routledge, 1993), 13.

71. Ibid., 13.

72. Cresswell, *In Place/Out of Place*.

The "Disappearance of Public Space": An Ecological Marxist and Lefebvrian Approach[1]

John Gulick

Responding to epochal changes in the social and spatial composition of the contemporary capitalist metropolis—novel varieties of public indigence, gentrification, surveillance, and segregation—urbanists of different stripes are writing books and staging conferences around a dramatic theme: "the disappearance of public space." Among the writings are Richard Sennett's *The Conscience of the Eye*, Elizabeth Wilson's *The Sphinx in the City*, Sharon Zukin's *Landscapes of Power*, Michael Sorkin's edited collection of essays—*Variations on a Theme Park*, and Nigel Thrift's "An Urban Impasse?". The conferences included "Remapping Public Space: The New Geographies" (held at Stanford University in mid-October 1993), "Bright Lights, Mean Streets: California as City" (held at the Oakland Museum in Oakland, California, from February 10–12, 1994), and "Cities on the Edge" (held in San Diego in May, 1994). The theme of "the disappearance of public space" draws commentators from a range of disciplines: architecture, comparative literature, geography, history, planning, and sociology.[2]

The Definitions

Whatever the forum—a singular author in dialogue with his/her imagined interlocutors, a book introduction tying together a compendium of essays, a plenary session at a conference—these commentators do not offer a coherent and consistent definition of "public space." This might lead one to conclude that all the commentators are remarking about the same phenomenon. In fact, they are not. Not one, but three overarching connotations of "public space" can be found: (1) physical property formally owned by the state, such as streets, parks, vacant government-owned buildings, and plazas, used by marginal social groups to fulfill basic subsistence needs; (2) those urban sign-systems that, according to semioticians and other specialists who decode relationships between representations and power, do not govern or surveil the desires and actions of subjects; and (3) the "public sphere" of settings where diverse citizens can see and be seen by one another, engage in enlightened political discourse, and encounter the bond of social solidarity. I will subsequently refer to these as the "public property," "semiotic," and "public sphere" definitions, respectively.

Each of these definitions contains two implicit normative moments: a grievance about what is undesirable about the alleged disappearance of public space and a prescription for remedying this loss.[3] The incommensurability of these definitions means there exists competing valuations about what is wrong with current urban life and also competing theories for the basic dynamics behind its degenerated condition that are not consciously recognized as such. The proponents of each definition do not acknowledge that they favor competing valuations or competing theories.

"Public property" centers on those physical spaces where the right of private persons (including that most powerful of legal "persons," corporations) to exclude other persons from "habitation" is suspended. Because in advanced capitalist society the state formally "owns" all physical spaces not privately held, this conception of "public space" refers to entities that are state properties. Hence, one facet of the "disappearance of public space" is the closure or sale of "public property"—libraries, beaches, playgrounds, and the like.[4] The "public property" definition must discriminate between public property that has been sold or rendered off-limits to all and that which has "disappeared" only for those marginal social groups—street vendors, sex workers, the homeless—who rely on it for their material maintenance. "Disappearing public space" also includes the state wielding its own legal monopoly on violence, e.g., to make public space "safe" for high-income shoppers, tourists, and downtown office workers, and, by implication, those segments of capital that benefit from this flow of commercial traffic. This definition of public space especially focuses on state property habitually utilized for basic subsistence purposes by those who, for a variety of such "factors" as race, citizenship, class, gender, and health, are excluded and often exclude themselves from private sites where wages are earned, meals consumed, and bodies rested.[5]

This definition is often omitted from scholarly discussion. At the plenary session of the Stanford conference, for example, when musing about "public space," not one panelist mentioned the homeless, the street peddlers, or the minority youth who face police harrassment when they "loiter" in consumption enclaves of the well-to-do. It took prompting from the audience to illuminate these issues. To be generous, the displacement of the "material resource" definition may be traceable to its "negative" quality. (On the other hand, it may be the result of an unyielding adherence to an idealist method). Those who use public space for subsistence do so generally because they lack the money that would secure permanent access to private means of consumption and not out of philosophic attachment to the virtues of conducting the most fundamental affairs of living in public settings. The indigent begging on public walkways or scavenging trash from public curbs hardly epitomize the adage, "city air makes men (*sic*) free." One might strongly disapprove of police harassment of the homeless, anti-panhandling ordinances, and other such draconian measures in a society marked by an absence of decent formal employment opportunities for the urban poor, welfare cuts, insufficient affordable housing, and the like. Yet this disapproval hardly amounts to a spirited defense of public space as a material resource or safety valve for those with nowhere else to go.

The second meaning ascribed to "public space" is "democratic semiotic space." One feature of a sizable portion of the built environment is that it may be sensually experienced and visually "consumed" without engaging in the formal process of commodity exchange (although increasingly the disappearance of physical public space impairs such activity). "Semiotic space" is any sign-system embedded in artifacts in the built landscape, the meanings of which are mediated by the "intentions" of landscape architects, the practices of landscape inhabitants, and the latent "systems" (such as a capitalist economy or a totalitarian state) that govern the landscape's development.

What makes semiotic space "democratic" and, hence, "public" is not the sensory accessibility of artifacts, but the assorted relationships between various architectural and design motifs deployed, their historical origin, their ideological effects, etc. Some examples of "democratic semiotic spaces" are built environments that feature aesthetic variety drawing on genuine local traditions and settings for viewing art and other cultural artifacts that do not require entry fees or do not promote one-sided interpretations canonizing private power. The more one-dimensional and ahistorical the meaning connoted by an artifact, the less "democratic" and "public" the semiotic space it encompasses is. Examples of this type of non-"public space" would be systems of meaning that circulate through gated residential communities with restrictive design covenants or through corporate art plazas.[6] An especially salient target would be the Rouse Company's restored waterfronts (e.g., South Street Seaport in lower Manhattan or Harbor Place in Baltimore) where "ahistorical" maritime motifs are allegedly used to stimulate nostalgia for seafaring adventure, a yearning that can "only" be resolved through purchase of a curio from one of the district's boutique tenants.[7]

One major flaw with the semiotic approach to public space is that it, as Henri Lefebvre says, "reduces space itself to the status of a message, and the inhabiting of it to the status of a reading. This evades both history and practice."[8] However subject to multiple, partial, and shifting interpretations, the built environment is not merely a text. It is also a concrete product of human beings collectively transforming nature under certain ecological, social, and cultural constraints, as well as a concrete medium enabling the transformation of these constraints.[9] Nonetheless, ascertaining meaning from the built landscape composes a "moment" in the overall project of making social space: too often structural Marxists have dismissed mental images as "epiphenomena," and ignored the fact that affective social individuals are the conduits of producing and reproducing sociospatial structure.[10]

However, the semiotic method rarely acknowledges that material experiences that generate interpretations of the built environment depend on where the subject is literally "situated"—on public turf or private turf, loitering or walking hurriedly to a late appointment, window-shopping or intent on a purchase—privileging instead the "decoding" of an expert theoretician. Lefebvre remarks that the specialist who discloses the ideological grammar of spatial sign-systems performs a specialized task nominally similar to that of the urban planner who charts traffic flows: both participate in the technocratic division of labor and the fragmentation of the lived, perceived, and conceived.[11] Not only do semiotic analyses apprehend significations from a synchronic point of view detached from concrete social individuals who carry out projects, but they also privilege impressionistic, visually oriented forms of knowing, reproducing both the splintering and the aestheticization of sensory experience pronounced in late bourgeois society.[12]

Even when entertaining how ideology inscribed in the built environment influences the practical activity of its builders and users, semiotic-oriented analyses confront a serious methodological problem. Very little empirical research is conducted, for example, to verify relationships between entwining consumers into various exotic sign-systems (such as the internal layout of a shopping mall, with all of its resplendent and bizarre juxtapositions of commodities) and actual purchases.[13] This is still a world in which people have basic physiological needs fulfillable through consuming historically specific contents; that capital accumulation restricts incomes imposes some degree of rational, utilitarian discipline upon the consumer when he/she calculates a hierarchy of possible purchases based on the commodity's price and its need-fulfilling promise. Semiotic-oriented analysis is stunted by the very epistemological problem that stunts the clients of advertisers themselves: when there is "technological diffusion" of the most sophisticated marketing techniques, it becomes impossible to tell how cognition translates into effective demand. But in other respects this epistemological quandary does not concern cultural theorists whatsoever, since the focal point is synchronic sign-systems and not material activity, surely a shortcoming for any analysis of public space with political aspirations.

The third meaning, which hearkens back to classical political theory, is the "public sphere," a milieu where members of the polity ("citizens") come together to engage in rational, constitutive political discourse and to experience sentiments of social solidarity.[14] The public sphere is both a site, such as a town square, and a process of republican self-governance, which goes on in a setting such as a town square. While analytically separable, the site and the process are comprehensible only with reference to one another, and the whole of the public sphere is greater than the sum of its parts.

Jürgen Habermas characterizes the public sphere as a milieu where self-reflexive and rational social individuals come together to fashion binding collective decisions, which govern different facets of society, such as the social division of labor or the allocation of various individual rights. John B. Thompson argues that Habermas's definition of the public sphere presumes that genuine "collective will formation" takes place in face-to-face encounters of discourse and deliberation. While Habermas cites the political tracts that ran in the literary journals of early modern Europe as constituents of the public sphere, the spatial context in which writing, reading, and discussion occurred, i.e., the salon or the coffee house, was of equal—if not more—significance to Habermas.[15] Physical proximity and visual transparency in the public sphere instill sentiments of social solidarity, in addition to being constitutive of rational political discourse. Seeing and being seen and witnessing social diversity enriches the civic character by fostering tolerance.[16] An example of this dimension of the public sphere is a flea market that brings the whole range of a polity's artisan and cultural groups physically together. A complex mix of celebrating communal existence and partaking in noninstrumental exchanges characterizes social fusion in this milieu, not the abstract, place-less, and instrumental logic of price-based transactions.[17]

Certain examples of a dissolving public sphere can be traced to socio-spatial restructuring in the advanced capitalist city, such as restrictions on distributing political literature in luxury hotel and shopping complexes, the slow death of independent bookstores where customers can "browse," or judicial rulings prohibiting controversial art in "public-private" facilities.[18] Other developments are much less directly connected, such as the privatization of information in commercial databanks. While miniaturization of information storage has reduced the floor-space needs of such a knowledge-intensive "industry" as insurance and hence made central city locations more affordable and attractive, the tightening of property boundaries around knowledge is not itself a direct outgrowth of "information industry" clustering in the core city.

The "public sphere" definition is downplayed in almost all of the literature at issue here. This lack indicates that those who fetishize material subsistence or semiotic readings tend to ignore concerns central to the flourishing of political community. A healthy, critical passion for exposing most assumptions of political agency as white, male, and bourgeois can ossify into a deconstructive fetish that denies the possibility of rational, constitutive political agency altogether. However, the public sphere definition implicitly recommends the maintenance of vibrant public spaces where diverse social groups may see and be seen by one another, thus rescuing the fiber of the *polis*. This panacea for urban ills marks the invisible relationship between the social classes as a central, if not the sole, source of urban conflict. This ethic slights the salience of class, status group, and racial divisions in structuring urban alienation and violence, and plays to the "illusion of transparency," the ruse that from sight comes knowledge and social harmony.[19] And as with all discourse dipped in the murky waters of what remains of U.S. liberalism, this ethic confuses "differences" that should be revered and protected (such as cultural/linguistic identity or sexual preference) with those that should be eliminated (most obviously, that of class).

While those commentators who espouse one particular definition of "public space" tend to ignore the other definitions (whether intentionally or unintentionally), and hence subscribe to narrow diagnoses of the "disappearance of public space," there are exceptions. Among the authors drawn to a "public property" definition of public space, those who focus on how actors navigate streets, plazas, and parks to evade the police, find a place to sleep, pick a site for panhandling, etc. acknowledge that meanings flow through the built environment. Otherwise, they would be suggesting that the homeless, the street vendors, the sex workers, etc. do not actively interpret their built surroundings in pursuing their subsistence strategies, but simply roam the streets, e.g., craving satisfaction of their prelinguistic instincts according to some Pavlovian logic. Reducing public space to a hunting ground creates the peril of reinforcing Social Darwinist opinions about the homeless and African-American teenagers.[20]

Mike Davis draws attention to the fundamentally semiotic aspect of urban design features that function to regulate the subsistence or leisure activity of urban marginals. An example would be the ramparts that cordon off the new core of "global city" Los Angeles. While physical access to the corporate downtown is not technically restricted, the imposing embankments "tell" the poor and minority youth that their presence will most likely be met with police hostility.[21] In general, though, the multiple discourses on the "disappearance of public space" do not explore how the scarcity of physical public property, the profusion of ideologically opaque built landscapes, and the demise of physical places that instill social solidarity may or may not be linked to one another.[22]

A Method for Sublating the Definitions

One way out of the maze of terminological differences, and hence the competing valuations, theories, and solutions that flow from these differences, is to elaborate a method that sublates the three definitions of the "disappearance of public space." A synthetic method would clarify how draconian panhandling statutes, "Afrikan" boutiques selling kente cloth to white middle-class tourists, and "virtual communities" of computer bulletin board addicts define one another, as opposed to being separate "happenings" that coincidentally share the same sociospatial setting of the advanced capitalist city. Such a method would explore the inner relations between the forced dismantling of tent cities in postquake Los Angeles, the invasion of fake adobe Taco Bells south of the U.S. border, and the absence of a vigorous free press in Mexico.

Such a method would derive connections not through some abstract declaration that each comprises a "moment" in a common structure (which explains nothing), but instead through postulating that human existence is the material and social product of the conscious, collective application of labor (using technology), to land and to nature, which then constrains and enables further collective activity. The earth and its living systems, and human beings and their systems of meaning, make up the organic bases of social life, whatever the social formation and its attendant class and status relations. Informed and coordinated by systems of meaning, or "culture," human beings—through socially organized, purposive activity—remake "first nature" and their own nature in tandem. However driven by class-rooted conflict or systems of social or imperial domination, human societies never fully alienate themselves from the organic bases of social life: land, labor, and culture.[23] The existing discourses on the "disappearance of public space" touch on questions of access to land as a means of subsistence, of the commodification of labor and alienation from communal political life, and of the degradation of systems of meaning, both those that provide a "sense of place" and those that provide sentiments of communal solidarity. But these discourses do so in a fragmented fashion, and without recognizing that land, labor, and culture are bound together as organic components of social existence. The ontological unity of human social existence, at however high a level of abstraction, dictates that a truly synthetic approach to the "disappearance of public space" entails sublation of the three definitions.

When the conditions of collective human existence are not only bought and sold on a market integrating a complex division of labor but also turned into conditions of capitalist production and accumulation, they are being used as capital.[24] If culture is defined in a Durkheimian sense, i.e., as a set of norms that binds members of a collective organization—be it a nation-state, a neighborhood, a business firm, or a social class—to treat and be treated by one another in a certain regular way, then it can be demonstrated that culture (i.e., a "mode of cooperation") may be valorized as a capitalist productive force.[25] For example, capital that takes advantage of a work force that is habitually deferential to management will reap more surplus value than capital that does not, *ceteris paribus*.[26]

Remaining in the Durkheimian vein, culture includes not only sets of beliefs manifested in moral precepts and routine social practices, but also systems of representation that express the principles that give cohesion to a social formation.[27] If, as those who prefer a semiotic approach to public space would have it, the built landscape can be read as a "text," then this "text" will express ideologically salient values. For example, the skyward thrust of a core city financial tower connotes the social power of money capital. One may reach this interpretation by one of several routes, depending partially on how esoteric one's knowledge is. In the dominant code of interpreting raw feelings, the sheer awe experienced in the presence of a high rise of massive scale may translate into a fetishistic reverence for the corporation whose logo is imprinted on the facade. The centrality of a downtown location may connote "being at the center of things"—being privy to exclusive information or "in the know." Or to the more learned interpreter possessed of a knowledge of urban land markets, a central location may signify the seeming economic and political clout of its tenant.

Obviously, not all members of a social totality recognize hegemonic representations, much less assign them the same meanings. Interpretations of the built environment vary along lines of class, gender, ethnicity, sexuality, and other pertinent sociological categories. The irreducibly contingent biographies of social individuals also shape landscape interpretation. Moreover, class, gender, ethnic, sexual, and individual identities are partially constructed through processes of decoding the built environment. Nonetheless, the point stands that for those who buy and sell ground space, cultural codes prevail that confer more normative worth upon certain places than others and allow the owners of those places to accrue more ground rent than they would otherwise accrue. Zukin contends that in the '70s and '80s firms would pay "extra" rent to gain prestigious Manhattan business addresses. William Cronon illustrates that land speculators in 1830s Chicago would manufacture hegemonic discourses about the "coming boom" and the honor and riches it would bestow to property owners in order to successfully inflate land values. In both of these cases, landlords accrued rent not only for ground and urban space they did not produce capitalistically, but also for an ideological code they did not produce explicitly, although they may have served to enhance it through promotional campaigns.[28] These examples underscore how any synthetic treatment of the "disappearance of public space" must embrace both the commodification and capitalization of both ground space/urban space and systems of "cultural authority" that designate specific locations, places, etc. as desirable or forbidding.

A synthetic approach to the "disappearance of public space" would then root itself in the foundational premises of the "organic bases of social life" outlined earlier. The next step would be to characterize the general tendencies in the contemporary social formation that facilitate and restrict access to physical space as a means of subsistence, encourage certain popular interpretations of the built environment to take precedence over others, and destroy sites where the collective being of the polity is affirmed. A proper synthetic approach will recognize that dynamic ecological systems ("nature") and intersubjective systems of meaning ("culture") are inextricable and universal foundations of any human society, but also that each of these "foundations" is modified by specific social forms of property entitlement and technological development. While the Polanyist and ecological Marxist precepts laid out earlier supply the general framework of the synthetic method, the "sociospatial dialecticians," most notably Henri Lefebvre, provide the specific theoretical instruments for looking at public space in the contemporary city. These instruments are provided, however, without categorical adherence to the concept of the "organic bases of social life," which is a necessary complement that needs to be continuously and vigilantly applied.

Lefebvre

The sociospatial dialecticians begin with the elementary proposition that space is not a mental apparatus that makes things-in-themselves apprehensible and manipulable, an empty stage that "contains" the human drama of social development, or an epiphenomenal "mapping" of the social relations of production. Space is both the "precondition and product, medium and result" of social life.[29] Lefebvre's notion of the "production of space" was forged by his concern that, in post-war France, state transportation, housing, and land use planners had compartmentalized and regimented working people's "spaces of everyday life." Where a fluid meshing of work, play, community association, and family reproduction spaces had existed before, a functionally segregrated network of office blocks, high-speed transit corridors, and nucleated apartments took its place. Under the aegis of the "neo-capitalist" state, alienation in the life-world of the worker ran parallel with real subsumption to capital at work.

Lefebvre did not celebrate the splintering of "everyday life" as the conquest of false ethnic or neighborhood consciousness. To do so would be to morally circumscribe political struggle against the capitalist state to exploitation in capitalist production and to deny that the quality of human interaction in all spheres of social life was worth fighting for. The planning agencies reproduce the social relations of capitalism not only by rationalizing the movements of commodities, workers, and consumers or by arranging cheap housing for proletarianized peasants and immigrants, but also by inducing manifold "splits" in popular consciousness. Sector-specific industrial districts keep workers in different industries from having the customary contact propitious for waging a general strike; neighborhoods deliberately split by national status foster ethnic chauvinism; the physical separation of the job from the household fragment concerns over wages and workers' control from concerns over gendered domestic roles and neighborhood well-being; the spatial isolation of hospitals, prisons, welfare agencies, and schools parcel out the various struggles against the bureaucratic state.[30]

The concept of the "production of space" registers only in the context of this drama. People occupy, produce, and sustain a social space through their daily routines of working, shopping, cooking, attending meetings, minding children, recreating, etc. The bonds of affinity or the anxieties of alienation lend certain artifacts and places encountered in social space a symbolic prominence, producing mental maps that may achieve expression in popular representational forms or in intimate individual topographies. Using neo-classical "public choice" economic theory and statistical modeling techniques, the planners classify the concrete activities of human beings into functional, temporal, and locational categories and then apply this abstract knowledge in the remaking of the city's spatial form to serve social order and capital accumulation.[31] Each part of this triadic structure possesses a clear and distinct meaning, definable only in relation to the meaning of the other two parts, but not reducible to them. Lefebvre's concern with the urban dimension of the struggle of everyday people against the capitalist state gives life to the general concept of the "production of space," although the triadic structure applies to milieus other than advanced capitalist cities.

The chief benefit of Lefebvre's concise terminology (and developing it in relation to a delimited context) is avoiding a reified, supra-meaning for the term "space." Given the necessarily ontological trappings of the word "space," an imprecise definition means everything and nothing at the same time. But a materialist notion of produced space may encompass more elements than those suggested by Lefebvre's triadic structure. One facet of capitalist space is land, which is bought and sold as a commodity, and hence has a price. It may be used as a means of production (appropriating raw materials for sale or for direct consumption) or as necessary ground space for other varieties of social activity, whereby ground rent accrues to the landlord.[32] Two general facets of socially-produced space that find peculiar expression under capitalist forces and relations of production are (1) "worked matter" or "second nature," i.e., the whole of physical material fashioned through the history of human engagement with "first nature" (including human bodies themselves) and (2) a set of relations between spatially disparate material activities, that produces various kinds of synergies or frictions and often demands mediation by the state.[33] Finally, there is Lefebvre's orientation to the terrain of everyday activities of sleeping, nurturing, working, eating, recreating, with accompanying individual and collective "mental maps" that highlight normatively valued and devalued places.

Lefebvre's scheme may also not devote sufficient attention to the ultimate dependence of everyday human life, and hence its trials and its resilience, on "first nature." Lefebvre acknowledges that there exists a nonreducible "physical space" of energy transactions that necessarily underlies both "social" and "mental space," but fails to recognize the dangers and contradictions of, as he puts it, the "ultimate voidance and destruction" of "first nature." For example, potable water is a form of "first nature" that is a basic building block of human subsistence and for which there is no known synthetic substitute. Present rates of capital accumulation and patterns of energy consumption (not to mention industrial poisoning, inadequate sewage systems, depletion of acquifers, and the ruination of watersheds) imperil the long-range supply of potable water and call into question whether the "voidance of first nature" and continued human existence are compatible. To be fair to Lefebvre, however, perhaps the most significant organic basis of urban social life of which productive capital takes advantage but does not pay for directly is human density—the very phenomenon that the planning agencies, through their scientific "representations of space," step in to rationalize and direct. A sensitivity to systems of meaning, where the social control designs of planners are symbolically illuminated in the built landscape and mediate individual and collective imaging of repression and freedom, complements Lefebvre's likely insight that modern public space is for traffic flow, not foraging or loitering. Before it can be successfully applied, a synthetic approach to the "disappearance of public space" must combine the insights of Polanyi about the "organic constituents of social life itself" and the insights of Lefebvre about the spatial form and practice of these organic constituents under advanced capitalism.

Marx

In working out a method to discover the inner logic of collective human existence, Marx introduced the abstraction "production in general"—the common-sense proposition that social labor is a "nature-imposed" necessity transcending all specific forms of producing and distributing a surplus. This proposition established the materialism of Marx's method, but, as Marx himself acknowledged, by itself it is a vacuous blanket statement. The dialectical "other" of "production in general" is the multitude of ways that actual social formations have organized or do organize the production of surplus and distribution of the means of concrete existence. The materialist method is made historical only by dialectically connecting the real abstraction of "production in general" with empirical analyses of the specific processes governing the allocation of social labor (how much time and energy are directed to specific activities and who performs them) and the distribution of its results (to what purposes the social product is put and how it is divided up amongst "final consumers").[34]

The dialectical method employed in the preceding manner must figure into a synthetic approach to the disappearance of public space. Ground space is a condition of production inherent in all social formations; the frictions or synergies of distance, i.e., spatial relations, necessarily enter into "production in general." The earth itself—land—is inextricably a transcendent feature of collective human existence and survival as we know it; systems of meaning expressed in built artifacts also characterize all modes of organizing the production and reproduction of social life. Finally, no social formation can function properly unless enough of its members feel that their social individuality is being affirmed in some way, whether this be satisfied through imbibing dominant ideology or through engaging in a common mission to transform society. But these general propositions ring hollow unless complemented by a concrete analysis of how capitalist forces and relations of production specifically (1) condition the usage of ground space and land (including its mineral contents and the organic ecosystems it "supports"), (2) structure relations of propinquity and distance, (3) configure semiotic codes objectified in the built landscape, and (4) regulate encounters that produce feelings of social solidarity.

Dialectical qualification does not end here. While there may be "general laws" of the capitalist structuring of space, specific bounded land-scapes always feature peculiar physical, natural, and cultural attributes that influence local ground rent structure, local usage of land as a means of production, and the predominance of local systems of meaning embedded in built artifacts. Local landscape differentiation within the totality of the global market is not just a product of the dynamics of capitalist accumulation and crisis. It always bears more than the trace of physical geography, natural resource endowments, and cultural traditions, as well as the specific history of places vis-a-vis the territorial expansion of the circuits of capital via primitive accumulation, imperialism, colonialism, neocolonialism, etc.[35] The concrete character of specific bounded land-scapes, in turn, become more intelligible if framed in dynamic interaction with the totality of the global division of capitalist production and the physical, chemical, and biological systems upon which it rests.

Conclusions

What lesson is thus to be learned for approaching the disappearance of public space synthetically and dialectically? One trait that all competing definitions—physical public property, semiotic codes, and places where communal solidarity is expressed—have in common is the advanced capi-talist metropolis, mainly the Anglo-American metropolis, as their "object of discourse." It is true that each particular metropolis possesses its own particular natural and cultural attributes that structure and are structured by a particular economic history in relation to the city's region, nation-state, and global sociospatial division of labor. Nevertheless, shared status as an advanced capitalist metropolis also counts for something. For exam-ple, in both New York City and Detroit, in order to acquire permanent housing, one must either own ground space or rent it from landed capital, unless one qualifies for a federal, state, or local housing subsidy. This implies that to be permanently housed one either owns productive prop-erty, sells his/her labor-power for a salary or wage, or receives a direct or indirect transfer engineered by the state.[36] Many Third World cities have very different patterns—informal shantytown settlements ring the me-tropolis and are tolerated by municipal and state authorities as a politi-cally acceptable solution to rural migration and urban housing shortages.

At the same time, the increasing usage of public property as a place to attend to the needs of the body and the increasing political battles regulat-ing access and monitoring behavior that almost inevitably follow have different immediate "causes" for different advanced capitalist metropo-lises. Denizens of a cardboard colony in a Manhattan subway tunnel face gentrification as the principal enemy, while squatters in vacant lots in

Detroit contend with deindustrialization.[37] In the United States, anyway, advanced capitalism may always produce the homeless, the poor, and the homeless poor, but the where and the why of this production will alter across time and space (and, hence, so will the forms and demands of movements). Shortages of affordable housing in New York City and manufacturing layoffs in Detroit may be internally linked by the political hegemony of U.S. finance capital in responding to economic crisis. But divergent varieties of displacement in Manhattan and Detroit are in part products of the divergent position of New York City and Detroit within the global and national hierarchy of capital flows.[38]

In order to avoid one-sided abstraction about the "disappearance of public space" in general, then, the object of investigation should be narrowed from advanced capitalist urban space to a specific genus of advanced capitalist urban space, namely, urban space in the so-called "global cities," which in the United States would certainly encompass New York City and Los Angeles, and to a lesser extent Miami, San Francisco, Chicago, Houston, Seattle, and Boston.[39] Beyond establishing a proper balance between theoretical abstraction and empirical observation, the advantage of honing in on global cities is that a great deal of recent urbanist research and writing has centered on the subject. But none of this research and writing has been conducted explicitly from an ecological Marxist or Lefebvrian epistemology offering a unified definition of "urban space" or "public space" characteristic of but not reducible to the global city. The realities of geographically disparate and socially incomplete development of capitalist production, capitalized nature, and capitalized culture demand that a synthetic approach to the disappearance of public space be "as radical as reality itself." In other words, the object of investigation should be delimited enough so as not to lead to the production of a knowledge that violates the site-specific character of particular advanced capitalist urban spaces. Theoretical categories are only as good as the dialectical reality they help to illuminate, and exercises in abstract theoretical "unification" are purposeless unless they assist in comprehending immediate and specific social and spatial problems. Applying an ecological Marxist and Lefebvrian epistemology would illuminate the crackdowns on the homeless, postmodern architecture, and the demise of noncommercial plazas: how all three of these types of "disappearing public space" are structurally bound together by the "sociospatial structure" of the global city. The familiar features of this "sociospatial structure" are the producer service high-rise clustering; the urban professionals working in these high-rises whose income is disproportionately spent on "personal services"; and the low-wage, mainly immigrant, personal service workers whose exploitation and superexploitation lowers the cost of reproducing urban professional labor-power. All are structurally bound to the global city's position within the global sociospatial division of labor.[40]

All three forms of "disappearing public space" take on extreme proportions in the global cities of contemporary world capitalism. In these "command and control centers" of a highly dispersed world economy, the realities and images of social polarization and spatial segregation are familiar and well established. Gleaming towers housing corporate or divisional headquarters and the financial, insurance, and telecommunications firms that serve them dominate the skyline of the historic city center. Middle managers and professionals who wish to live in luxury apartments, condominiums, and restored row houses proximate to their places of work in the new transnational corporate downtowns put upward pressure on land rents and skew the housing market toward high-income groups. These enclaves of upscale work, residence, and entertainment must be sealed off or at least safeguarded from the unpleasant and potentially disruptive presence of those other global city dwellers whose labor and housing market situation is not so favorable. Among these other inhabitants of the global city are oppressed minorities who face, on the one hand, the destruction of their long-standing neighborhoods by gentrification, and, on the other, worsening job opportunities effected by the deindustrialization and restructuring of the urban economy. Recent migrants fleeing the poverty and misery of the Third World (for lack of a better term) comprise another distinctive stratum of the global city. For many, if not all, of these recent migrant groups, the only occupational niches to be carved out are in the informal sector, where work is often performed in public view. Another more privileged stream of migrants continuously flows into the global city—young, single adults, but especially the college-educated and cosmopolitan young. They are drawn to the global city for a host of reasons: to experiment with or practice innovative life-styles and to imbibe the cultural vitality generated by the city's diverse populations (in however commodified a fashion), but also to witness the spectacle of grotesque contrasts and latent violence in such a polarized setting. Cynical Generation Xers are magnetized by the "combat zone" atmosphere of the global city (however safe from it they may be). Myriad branches of the culture industry—from fashion magazines to record labels to music clubs—mine this semiotic construction of the global city in order to sell their wares. In the global city, the increased surveillance of public property, the almost instantaneous commodification of various "meanings of the city," and the loss of social solidarity are all tied together. Given the inequitable social composition of the global city and its pivotal connection to the world capitalist social division of labor, the "disappearance of public space" in all senses of the phrase is no accident—here, it takes on its sharpest and most profound expressions.

Notes

1. This paper was originally written and submitted as a master's essay to the Sociology Board of Studies, University of California, Santa Cruz, May 1994. The author would like to thank the following individuals for formal and informal remarks and criticisms that assisted in the preparation of this paper: James O'Connor, Wally Goldfrank, Sharad Chari, Patricia Allen, Michael Johns, Robert Connell, Alan Rudy, Robert Meister, Will Hull, Raegen Rasnic, and the anonymous reviewers of *Philosophy and Geography*.

2. Richard Sennett, *The Conscience of the Eye* (New York: Knopf, 1990); Elizabeth Wilson, *The Sphinx in the City: Urban Life, the Control of Disorder, and Women* (London: Virago Press, 1991); Sharon Zukin, *Landscapes of Power: From Detroit to Disney World* (Berkeley, Calif.: University of California Press, 1991); Michael Sorkin, editor, *Variations on a Theme Park: the New American City and the End of Public Space* (New York: Noonday Press, 1992). In *The City Builders* (Oxford and Cambridge, Mass.: Basil Blackwell, 1994), Susan Fainstein supplies a list of those authors who offer what she (inaccurately) calls "post-structuralist" critiques of the exclusionary and inauthentic elements of contemporary urban design, as well as a critique of their critiques. Fainstein provocatively challenges the "post-structuralists" to show that the contemporary capitalist metropolis is any more segregated or socially controlled than its predecessors. She also appropriately contends that there is nothing "ingenuine" about restoring run-down warehouses as glorified shopping malls. On the contrary, she sees such reuses of dilapidated industrial and commercial infrastructure as a most authentic manifestation of a consumer capitalist order. Nigel Thrift confirms that there has been a recent deluge of writing and talking about a seemingly new advanced capitalist urban order, including a changed shape and function of "public space" within this new order. While I agree that "analyses of the contemporary Western city have become familiar, even predictable, circling around and worrying the same issues to increasingly little effect," I disagree with Thrift's attempt to provoke new discourse through post-modernist analytic categories—which, in my mind, already characterize much of the literature and discussion to a fault. Instead, I propose an ecological Marxist and Lefebvrian approach, outlined and detailed later in this paper. See Nigel Thrift, "An Urban Impasse?" *Theory, Culture and Society*, 10 (1993): 229–38.

3. This parallels the dialectical characteristic of any "objective" or "factual" statement about the world issued by a historically situated, existential subject. In any descriptive utterance there are always traces of (1) theorizing the structural parameters and historical forces that make the desirable phenomenon possible, and (2) projecting a vision of what it means to be "fully human" in such a material context and taking on responsibility for pursuing the fulfillment of this vision. I borrow this notion of the dialectic between description and prescription, analysis and valuation, from Sartre in *The Critique of Dialectical Reason*. See Mark Poster, *Sartre's Marxism* (London: Pluto Press, 1979), 22–23, 36–37, and 51–52.

4. On beaches, see Marilyn Gardner, "The not so wide open spaces," *Christian Science Monitor*, 23 July 1991; on libraries, see Marilyn Gardner, "Symbolic

ways to burn a book," *Christian Science Monitor*, 16 April 1991; on public parks, see editorial, "Park prospects," *New Yorker*, 14 February 1994, 6–7; on all varieties of state property, see Victor Kirk, "Fiscal fire sale," *Fortune*, 27 June 1992.

5. In the aforementioned Sorkin volume, Mike Davis and Neil Smith both put forward narratives that stress public space as a physical resource that underprivileged social actors use to fulfill their material and cultural needs. See Mike Davis, "Fortress Los Angeles: The Militarization of Public Space," in *Variations on a Theme Park*, ed. Michael Sorkin,154–80; Neil Smith, "New City, New Frontier," in *Variations on a Theme Park*, ed. Michael Sorkin, 61–93.

6. At the aforementioned Stanford conference, Michael Brill furnished both the "positive" and the "negative" examples.

7. For a semiotic critique of the Rouse Company's restored waterfronts, see M. Christine Boyer, "Cities for Sale: Merchandising History at South Street Seaport," in *Variations on a Theme Park*, ed. Michael Sorkin, 199–204.

8. Henri Lefebvre, *The Production of Space*, trans. Donald Nicholson-Smith (Oxford and Cambridge, Mass.: Basil Blackwell, 1991).

9. As another French commentator, Raymond Ledrut, says, "The city acts as a text only analogically and is, in reality, a pseudo-text. Similarly the city acts only analogically as a sender and is, in reality, a pseudo-sender . . . The production of the city is a social practice, even if it involves groups and classes some of which have more power and influence than others." Raymond Ledrut, "Speech and the Silence of the City," in *The City and the Sign*, ed. Mark Gottdiener and Alexandros Lagopoulos (New York: Columbia University Press, 1986), 114–34. To quote philosopher Kate Soper, "The fact . . . that builders cannot build without a knowledge of the language of building materials does not mean that slabs and blocks have the ontological properties of words, or what they construct is a house of discourse." Kate Soper, *Troubled Pleasures* (London: Verso Press, 1990).

10. Stanley Aronowitz, in his *Science as Power: Discourse and Ideology in Modern Society* (Minneapolis: University of Minnesota Press, 1988), notes the family resemblances of both structuralism and semiotics —their synchronic emphases and rejection in the name of "science" of inspecting the "lived experience" of bounded agents. Pierre Bourdieu, in his *Outline of a Theory of Practice* (Cambridge: Cambridge University Press, 1972), attacks structuralism for its atemporal modeling, which reflects the peculiar reconstructive epistemology of the ethnographer and is not faithful to the strategic action of agents. John Jakle and David Wilson, in *Derelict Landscapes: The Wasting of America's Built Environment* (Savage, Md.: Rowman & Littlefield, 1992), argue for a method that focuses on individuals as "landscape degraders," although not a method that abstracts away individual property resources, class location, or ideology. R. S. Philips, in his "The Language of Images in Geography," *Progress in Human Geography*, June 1993, 180–94, contends that mental images should be assigned the same ontological status as physical landscapes or the maps of the cartographers.

11. Lefebvre, *The Production of Space*.

12. For corroboration, see John Fiske, "Cultural Studies and the Culture of Everyday Life," in *Cultural Studies*, ed. Lawrence Grossberg, et al. (New York: Routledge/Chapman/Hall, 1992), 154–73, and also Philips, "The Language of

Images in Geography." Unfortunately, the principal countertendency to semiotic approaches in social theory is a naive celebration of "consumer resistance" and "transubstantiation of meaning" in processes of commodity contemplation, purchase, and display. The problem here is not so much assigning excessive meaning-constituting powers to the consumer, as it is stressing T.V. shows, fashion, music, etc. as exigent matters of political struggle.

13. Rudi Laermans, "Bringing the Consumer Back In," *Theory, Culture, and Society* 10 (1993): 153–61.

14. Among those who have extended the concerns of classical political theory into diagnoses of post-World War II advanced capitalist societies, the most notable exponent of this definition of the "public sphere" is Hannah Arendt. See Arendt, *The Human Condition* (Chicago: University of Chicago Press, 1958).

15. According to Thompson, Habermas thus forecloses the possibility that modern electronic media may be used to facilitate rational communication; contemporary technologies are predisposed to engineering consent. While this may or may not be an accurate interpretation of Habermas, it is still useful to *analytically* distinguish the "process" from the "site" notion of public space, if only to explore the possibility that physical proximity is not a necessary precondition for "undistorted communication." See Habermas, *Structural Transformation of the Public Sphere: An Inquiry into a Category of Bourgeois Society* (Cambridge, Mass.: MIT Press, 1989); John B. Thompson, "The Theory of the Public Sphere," *Theory, Culture, and Society* 10 (1993): 173–89. For another interesting argument on how electronic media (specifically electronic mail networks) are inherently destructive of the personal ontology necessary for rational thought, see Sven Birkerts, "The Electronic Hive: Refuse It," *Harper's*, May 1994, 17–20.

16. Trevor Boddy is implicitly endorsing a "public sphere" definition in these remarks: "The reversal in the 1980's of two decades of improved urban amenity will only increase pressure for the construction of the analogous city, because too many of us—especially at the highest level of corporate and political life—have lost faith in the possibility of a socially diverse, multiracial, tolerant, public urban realm . . . We should not sacrifice the life of the polis, that most ancient benefit of the culture of western cities, with meek excuses about private property rights and the desires of people to associate with their own kind." See Boddy, "Underground and Overhead: Building the Analogous City," in *Variations on a Theme Park*, ed. Michael Sorkin, 123–53.

17. This example is drawn from Michael Brill at the Stanford conference.

18. On the stifling of First Amendment rights in shopping malls and electronic space, see Herbert I. Schiller, *Culture Inc.: The Corporate Takeover of Public Expression* (New York: Oxford University Press, 1989); on gentrification and independent bookstores, see Russell Jacoby, *The Last Intellectuals: American Culture in the Age of Academe* (New York: Basic Books, 1987); on courts declaring federally owned, administered, and subsidized train stations devoid of free speech protections, see David Cole, "In Your Space," *The Nation*, 14 March 1994, 329–30.

19. "The illusion of transparency goes hand in hand with a view of space as innocent, as free of traps or secret places . . . Such are the assumptions of an

ideology which, in positing the transparency of space, identifies knowledge, information, and communication. It was on the basis of this ideology that people believed for quite a time that a revolutionary social transformation could be brought about by means of communication alone . . . Such agendas succeed only in conflating revolution and transparency." From Lefebvre, *The Production of Space*.

20. Mark Gottdiener and Alexandros Lagopoulos characterize "cognitive mapping" as the dominant approach to how actors apprehend and interpret the built environment. In this mainstream approach, firmly rooted in behavioral psychology, the actor is an abstract individual possessed of universal perceptual equipment, whose activity is guided by processing and responding to paths, lines, nodes, and other morphological features. This model ignores the social class or political status of the subject and the relationship between its concrete "project" and its apprehension of its built surroundings and privileges perception of raw data over conception of ideological artifacts, given that the latter is not observable in accordance to positivist methods. Human subjectivity is thus reduced to that of a laboratory rat wandering through a maze—and behaviorism is notorious for underestimating "nature's subjectivity" as well. See *The City and the Sign*, ed. Gottdiener and Lagopoulos (New York: Columbia University Press, 1986).

21. Mike Davis provides an extensive analysis of the social control function of urban design in contemporary Los Angeles in *City of Quartz* (London, New York: Verso Press, 1990).

22. The landmark exception to this enclave mentality is David Harvey's *tour de force, The Condition of Postmodernity*, which in its totalizing scope may have spawned a counterreaction of increased insularity, of which Thrift's article is representative. Zukin also connects the rise of "post-modern" cultural landscapes in the United States to its incorporation into the global circuit of production and consumption. See Harvey, *The Condition of Postmodernity* (Oxford and Cambridge, Mass.: Basil Blackwell, 1989); Thrift, "An Urban Impasse?'; Zukin, *Landscapes of Power*.

23. The economic anthropologist Karl Polanyi introduces the notion of the organic preconditions of human social existence: land, or nature, and labor, or cooperative human activity producing subsistence and building civilization. For Polanyi, "culture" is a secondary term denoting the harmonious, if not egalitarian, regulation of the nature-society metabolism, prior to the imposition of market society, which "disembeds" land use and labor allocation from norms of cultural reproduction. See Polanyi, *The Great Transformation* (Boston: Beacon Press, 1944).

24. In his landmark essay, "Capitalism, Nature, Socialism: A Theoretical Introduction," James O'Connor distinguishes the *commodification* of land, labor, and culture—that is, their status as entities to be bought and sold on capitalist markets—from the *capitalization* of land, labor, and culture—that is, the tapping of their use-values in capitalist production processes to produce surplus value. See O'Connor, "Capitalism, Nature, Socialism: A Theoretical Introduction," *Capitalism, Nature, Socialism*, Fall 1988.

25. Of Emile Durkheim's various and sundry works, the one that most informs my understanding of a distinctly Durkeimian approach to culture as a superindi-

vidual set of meanings that enable human social conduct is *The Elementary Forms of the Religious Life* (New York: The Free Press, 1915).

26. Again, I borrow this example of how an organic constituent of social life can be twisted into a productive force for capital from O'Connor. See O'Connor, "Culture, Nature, and the Materialist Conception of History," manuscript, 1993.

27. While I advocate borrowing Durkheim's notion that individual space/time sensibilities are ordered collectively and may manifest themselves in various collective narrative forms, I reject Durkheim's neo-Kantian (and anthropocentric) idealism, whereby the world of non-human nature is an unintelligible, undifferentiated mass save for its disclosure by the collective categories of understanding.

28. Zukin, *Landscapes of Power*; William Cronon, *Nature's Metropolis: Chicago and the Great West* (New York: Norton & Company, 1991).

29. The words here are Edward Soja's, from Soja, "The Spatiality of Social Life: Towards a Transformative Retheorization," in *Social Relations and Spatial Structure*, ed. Derek Gregory and John Urry (London: MacMillan Press, 1984).

30. I draw this characterization of the relation between Lefebvre's political position and his definition of "space" from Gottdiener in both *The Social Production of Urban Space* (Austin, Tex.: University of Texas Press, 1985), and "Debate on the Theory of Space: Toward an Urban Praxis," in *The Capitalist City: Global Restructuring and Community Politics*, ed. Michael Peter Smith and Joe Feagin (Oxford and Cambridge, Mass.: Basil Blackwell, 1987), 199–218.

31. In *The Production of Space* Lefebvre characterizes this triad of "produced spaces" as "spatial practice," "representational spaces," and "representations of space."

32. Harvey has written extensively on the role of ground rent in capitalist societies, especially on the topic of how opportunities for real estate speculation temper the rhythms of capitalist growth and create rifts between different fractions of capital. See Harvey, "Land Rent Under Capitalism," *The Urban Experience* (Baltimore: Johns Hopkins Press, 1989).

33. Neil Smith offers a hard-line materialist and constructionist approach to the issue of "second nature." Citing (a) the capitalist appropriation and application of scientific knowledge and (b) the spatial expansion of capitalism to encompass the whole globe by the late nineteenth century, Smith seems to argue that not only is the "nature" encountered in the twentieth century strictly a "second nature," but, moreover, a capitalistically produced "second nature." Smith acknowledges that world capitalism is a geographically uneven system, with different types of "second nature"-making and "second nature"-transforming collective activities prevailing in different locales. Hence, "second nature" itself is not a homogenous totality, uniformly spread across the whole world, but differs in local form and content. Yet the local form and content of "second nature" depends wholly on where that locality fits into the spatial hierarchy of the world wide capitalist social division of labor. While the standing of localities within this hierarchy may change over time, the shifting fortune of localities is ultimately governed by the dynamic and contradictory processes of innovation and disinvestment innate to world capitalism. Smith thus loses sight of the extent to which local "natural conditions of production" (climate, topography, geology, flora and fauna, and so

on) influence why and how places are incorporated into the circuits of world capitalism, and the ways in which these local environmental features, however transformed, have a certain durability and continue to shape the "second nature" of a place. See Smith, *Uneven Development: Nature, Capital, and the Production of Space* (Oxford and Cambridge, Mass.: Basil Blackwell, 1984).

34. Among others who have written intelligently about Marx's method, Derek Sayer offers helpful insights about how best to move back and forth between Marxist theoretical constructs (such as "forces of production" and "relations of production") and really existing class-dominated production processes that both flesh out and inform these constructs. See Sayer, *The Violence of Abstraction* (Oxford and Cambridge, Mass.: Basil Blackwell, 1987).

35. See Alan Pred, "Sounds and Silences, or the Author's Voice and Voices: A Commentary on William Cronon's *Nature's Metropolis*," *Antipode* 26:2 (1994), 147–51.

36. See Shoukry Roweis, "Urban Planning in Early and Late Capitalist Societies: Outline of a Theoretical Perspective," in *Urbanization and Urban Planning in Capitalist Society*, ed. Michael Dear and Allen J. Scott (New York: Methuen, 1981), 159–78.

37. For verification, see Jakle and Wilson, *Derelict Landscapes*. Thanks also to Doug Henwood, personal communication, 23 November 1993.

38. Michael Peter Smith and Joe Feagin, "Cities and the New International Division of Labor: An Overview," in *The Capitalist City: Global Restructuring and Community Politics*, ed. Smith and Feagin (Oxford and Cambridge, Mass.: Basil Blackwell, 1987).

39. Both Norman Glickman and Saskia Sassen offer their separate criteria for what makes a city a "global city" or a "world city." See Norman Glickman, "Cities and the International Division of Labor," in *The Capitalist City: Global Restructuring and Community Politics*, ed. Smith and Feagin, 66–86, and Saskia Sassen, *The Global City: New York, London, Tokyo* (Princeton, N.J.: Princeton University Press, 1991).

40. See Sassen, *The Global City* and, Manuel Castells, *The Informational City* (Oxford and Cambridge, Mass.: Basil Blackwell, 1989).

Contested Space: The Rural Idyll and Competing Notions of the Good Society in the U.K.

Matthew Gorton, John White, and Ian Chaston

The 1980s and 1990s have witnessed growing demands on, and competing claims to, the use of rural space. In attempting to support their own claims, proponents often invoke notions of what the rural should be—the rural idyll. This paper considers the competing (and often mutually exclusive) notions of individualist and communitarian justice attached to the nebulous concept of the rural idyll and seeks to demonstrate that processes of diversification and fragmentation in rural localities have intensified these conflicts as to what should be rural. The next section considers the reordering processes of rural economic diversification, counterurbanization and agricultural restructuring and their promotion of inter- and intra-rural variations. The attempts of researchers to capture and order this diversity is discussed in the proceeding section together with a critique of the orthodox approaches taken. It is argued that customary sources of power have been diluted and new forms evolved so that the policy locality is more heterogeneous and less shapeable by traditional and individual actors. This has led directly to more diverse and mutually exclusive normative arguments concerning the nature of rurality.

Contested Space: Reordering Rural Britain

Three broad processes of restructuring can be outlined with regard to rural areas in the U.K.: counterurbanization, economic diversification, and the declining socio-economic and political importance of agriculture. These processes have led to more diverse and fragmented rural localities. Historically, rural communities were not homogenous and harmonious societies—far from it; but, traditional dichotomies aimed at understanding rural communities, particularly the landed versus landless binary classification[1], are no longer appropriate, if they ever were.

Counterurbanization

Counterurbanization can be defined as a deconcentration of population from urban areas due to net migration flows into rural localities. This process can be divided into two distinct forms: that within the urban-rural fringe and that arising in remote rural areas.[2] The first sizable migrations out of urban areas, which began during the interwar era were to rural areas immediately surrounding conurbations, particularly around London.[3] Until about the mid-1960s growth was most extensive in the accessible countryside, within easy commuting distance of major employment centers. Since the 1970s, migration has spread further out to remoter regions reversing long term trends of depopulation.[4] Some of these gains have been substantial—in the county of Cornwall, for example, between 1971 and 1981 there was an increase of fifty thousand on a base population of three hundred eighty thousand.[5] For Britain as a whole between 1971 and 1981 the population of rural areas rose by nearly 10 percent, while the total population of England and Wales increased by only 0.5 percent.[6]

Following Sant and Simons, counterurbanization can be seen as having three major dimensions:

1. *Place Utility*: the value put on different living environments.
2. *Ability to Move*: the economic and personal attachments one has to present locations.
3. *A Willingness to Move*: the desire to relocate to nonmetropolitan localities.[7]

Significant numbers of the population, particularly the middle classes, have consistently placed a premium on rural living environments (the first dimension), as opposed to urban ones. This rural idyll initially became obtainable through the growth of private transport that facilitated commuting and the provision of reasonably cheap housing in the immediate postwar era (the second dimension).[8] More recently, further advances in transport infrastructure and improved communications have meant that urban dwellers have been able to move further from the metropolitan core and maintain the degree of mobility and service provision they desire (the third dimension). In addition, tight planning controls, i.e., green belt legislation, around the major cities have pushed housing pressures further out into freestanding towns and villages and more remote regions.[9]

Boyle examines the characteristics of migrants moving into remote rural areas derived from the Sample of Anonymised Records (SARs), which is provided as part of the output from the 1991 U.K. census.[10] The SARs provide information on migrants' social and economic characteristics and their housing circumstances. As part of the study the author compares the characteristics of in-migrants to rural Wales and Scotland (3,278 records) with the profile for "natives" (11,557 records). The migrants were more likely to be aged less than 20, between 40 and 49, and over 65, divorced, from an ethnic minority, and self-employed. Negative parameters included public and private-rented accommodation, and having been widowed.

More generally, the migrants were more highly qualified and represented among the service class and the armed forces, much less likely to be employed in agriculture or construction, but more likely to be self-employed without employees, unemployed, retired, or other inactive. Those coming in are thus also more likely to be middle class, economically inactive, or unemployed. As Boyle notes:

> . . . it appears that these newcomers are more likely to be over-represented at either end of the social spectrum with both groups causing tension among the more "average" Welsh and Scottish. Thus, they may either be more privileged and consequently regarded jealously, or less active and therefore regarded as being a burden.[11]

The process of counterurbanization and population growth within rural areas cannot be overgeneralized and as a reordering phenomenon it must be seen as promoting heterogeneity rather than a new form of homogeneity. The process of counterurbanization has led to the influx of new groups to rural areas with migrants forming a diverse economic and social spectrum. Moreover, it is argued that, by forming such a wide array and using the terms of Sant and Simon's place utility, the way in which migrants value the countryside is multifarious and cannot be attributed to a single disposition.[12]

Agricultural Restructuring

U.K. agricultural production has risen extensively in the postwar period, first under a system of price supports stemming from the 1947 Agriculture Act and then, since 1973, under the Common Agricultural Policy (CAP) of the European Union (EU). Britain has gone from an interwar position of being able to produce only one-third of its domestic food requirements to being a significant net exporter of temperate food stuffs.[13] These rises in agricultural production have been achieved by increased mechanization, input intensification, and farm amalgamation with consistent falls in the number employed; within Britain since 1951, the number of farms has fallen by about one half, and the work force has been reduced by approximately three quarters.[14] Between 1973 and 1983 in the U.K. (the first decade of EU membership), there was, at constant prices, a 49 percent increase in gross product per person engaged in agriculture while during the same period the agricultural labor force declined by 12 percent.[15] For the three key commodities of wheat, barley, and milk, for example, the increases in yield for the U.K. between 1973 and 1983 were 45 percent for wheat per hectare, 17 percent per hectare for barley, and 25 percent for milk per cow.[16]

By the beginning of the 1980s even within rural areas the percentage of the work force engaged in agriculture (14 percent in 1981) was half that in manufacturing and construction (28 percent), and less than one third of that engaged in services (58 percent).[17] Capital intensification accelerated during the 1980s with farming playing an increasingly subordinate role within highly complex agri-food chains, with an ever diminishing role for labor.[18] U.K. agriculture in 1993 accounted for only 1.4 percent of the gross domestic product (GDP) (the lowest within the EU) and only 2.1 percent of total employment (also the lowest within the EU). Measured in terms of contribution to GDP, agriculture is, in fact, a less important part of the agri-food than food and drink manufacturing, distribution, and catering as their corresponding figures are 2.9, 2.7 and 1.9 percent of GDP respectively.[19] Even in the most rural counties of the U.K. the percentage of the working population engaged in agriculture, forestry, and fishing does not exceed 15 percent. Britain's rural economy can no longer be seen as agrarian-based and, thus, the space taken up by agriculture is in stark contrast to the employment opportunities it now offers.

This implies that agriculture and related industries will not provide nearly enough jobs for the rural work force and there is a clear need for diversification within rural economies. This diversification has been achieved in many areas via market-led changes, such as counterurbanization and most of the urban-rural shift in manufacturing and services. However, in peripheral and less attractive areas, the extent of these processes has been more limited. These peripheral localities have not attracted a large manufacturing base and thus tend to be characterized by a low-skilled, poor-opportunity local economy of ever-decreasing agricultural demand and a service sector highly geared to part-time and seasonal work, particularly in tourism, catering, and retail sectors.[20]

This loss of agricultural hegemony has been reflected in the changing nature of rural political systems. At a national level, there has been a clear challenge to the enclosed "holy trinity" policy community of MAFF with its "single minded commitment to the farmers cause," the NFU (National Farmers Union), and the CLA (Country Landowners' Association).[21] A shift in emphasis has occurred away from primary food production to consumer rights and safety, with a widening of the environmental agenda from a conservationist view, concentrating on visual amenity and habitat protection to a less anthropocentric view and the assertion of tentative public environmental rights. At a supranational level, CAP reform has been linked to the notion of a polyvalent rural employment structure, that recognizes the need to manage the transition to an economic and social system in which farming is not the major employer of the rural workforce.[22] At the national level there is thus a clear understanding that the welfare of agricultural interests and the well-being of the rural community as a whole are more and more detached. In short, agricultural policy can no longer be seen (if it ever should have been) as a synonym for rural policy. The primacy of farmers in rural decision making has been challenged so that:

> Not only are there more bodies wishing to be consulted, but there is a much greater awareness of the interaction of policy changes in the various sectors . . . while there is still an expectation that the ministry will "fight the agricultural corner," civil servants are now much more cautious in going in a producer-led direction. Interviewees in MAFF suggested that the department has lost any stomach it ever had to defend agriculture in an unreserved manner. These broad pressures lead us to query whether characterisations of the agricultural policy process which emphasise exclusion are still appropriate.[23]

Diversification of the Economic Base

Symbiotic with counterurbanization has been a diversification of the rural economic base. Using Census of Employment data, Hodge and Monk show that since 1970 rural areas are the only category of localities to have experienced a net increase in manufacturing employment.[24] The actors responsible for this diversification of the rural economy have come principally from outside the agricultural community and encompass a wide range of economic units from small businesses to multinational corporations. Rural small business growth has been stimulated by rising local demand that has emerged from net migration flows.[25] Large plants have relocated from inner-city locations to greenfield sites for a variety of reasons: lower land costs, greater accessibility, cheaper labor costs, higher labor productivity, and tax incentives.[26] Meanwhile farmers have not substantially diversified despite national and supranational level grants, advice, and training aimed at stimulating alternative enterprise creation. The majority of farmers have remained stubbornly committed to food production (aided by the CAP) and have dealt with periods of low profitability by reducing input costs (especially labor) rather than radically restructuring their businesses.[27] Townroe and Mallalieu, analyzing the motivations and types of new firm founders in rural areas, found that only 6 percent of new businesses created could be attributed to diversification by farmers.[28] This implies that not only has the importance of agriculture to the rural economy been diluted, but also the importance of farmers as economic actors has been diluted.

While usually a local phenomenon, studies of the establishment of small businesses in rural areas show the long-term importance of counterurbanization. Keeble and his colleagues, in their study for the Department of the Environment, consider the origins of founders of small businesses in accessible rural, remote rural and urban localities (see table 9.1).[29] The fact that over one-fifth of remote rural owner/managing directors (OMDs) moved there to set up their firms provides an unequivocally strong connection between population migration to rural areas and subsequent or contemporaneous new enterprise formation in these localities. The importance of migration, followed discontinuously by the establishment of new enterprises in rural areas, is also highlighted and this is a very different profile from urban founders who are predominantly indigenous.

Two key findings can thus be drawn out. First, the effects of these reordering processes has been to increase historically prevalent intra- and inter-rural variations in income. This is clearest when comparing the contrasting labor market fortunes of rural counties. For example, in April 1991 outside Greater London, the rural core counties of Southern England had the highest average male gross full-time earnings (£372.80 per week [p.w.] in Berkshire and £361.00 p.w. in Surrey) while the rural peripheral counties had the lowest (£246.30 p.w. in Cornwall, £260.20 p.w. in Dyfed, and £268.10 in Devon).[30] A similar picture emerges from com-

TABLE 9.1
Origins of Small Business Founders in Differing Localities

| Founders | Company Location | | |
	Remote Rural (%)	Accessible Rural (%)	Urban (%)
Born in the County	42.4	34.2	65.6
Moved to the County Before Setting Up Firm	36.5	52.5	25.9
Moved to Set up the Firm	21.1	13.3	8.6
TOTAL	100.0	100.0	100.0

Source: Keeble, P. Tyler, G. Broom, and J. Lewis, *Business Success in the Countryside: The Performance of Rural Enterprise* (London: HMSO, PA Cambridge Economic Consultants Ltd. for the Department of the Environment, 1992), 14.

parative female figures with the lowest average figures being £178.00 p.w in Cornwall and £174.58 in Borders.[31] The degree of income difference is thus now much greater between rural areas than between core and peripheral urban counties. Second, the increasing diversity and fragmentation of rural communities has meant the dissipation of traditional rural aristocracy and gentry hegemony and their control over local politics, culture, and economy. The dominance of agricultural interests and actors has long gone—with the new economic protagonists in rural areas coming from outside this community. This means that understanding the way in which individuals and groups appeal to each other and the claims to justice they make are vital to our comprehension of rural communities.

Diversity and Contested Space: The Need to Understand the Normative Dimension

Contemporary commentators have rejected Frankenberg's morphological continuum between rural and urban modes of living in that it ignores the diversity and interpenetration between, and within, the modes.[32] Sociological characteristics of a place cannot simply be "read off" from its relative location on the continuum.[33] Young and Wilmott discovered, for example, "model rural societies" in the East End of London while Connell and Pahl found the foundations of "urban" societies in central Surrey and Hertfordshire respectively.[34] The structuralist-functionist approach similarly failed to capture the heterogeneity of existence within the range of localities—a weakness that, since Pahl, researchers have sought to ameliorate. Successive commentators have sought to identify and classify the various groupings (stemming from the heterogeneity of existence) that are present within rural communities and the conflicts between them. The most ubiquitous and basic classification employed has been the disjunction between newcomers and locals.[35] This approach is typified in the quotation below from Strak:

> Conflict often arises between the new rural dwellers and established ones. . . . The later group wishing to see more economic growth and employment from new industry and commercial development. The former group to preserve the countryside and refuse any development.[36]

Such a scheme, however, does not allow for differences between "locals" and between "newcomers" and sees each group with a constant and predetermined set of preferences living in separate, bounded worlds. As the empirical evidence from Boyle details, newcomers and locals must be seen as heterogeneous clusters—the realization of which has led to the promulgation of more complex classifications.[37] This has led to more and more detailed classifications. For example Robinson identifies six groups of nonretirement migrants in Cornwall alone:

(1) Career transients or "spiralists"
(2) Commuters
(3) Professional self-employed and small business owners
(4) Small manufacturers, people seeking a "new start," and people involved in permanent job transfers
(5) Job-specific and nonspecific migrants
(6) Partial employment migrants: those choosing to be employed for part of the year (normally in the tourist industry)[38]

This approach does not explain, however, how such diverse groups exist within the same space/time continuum, and it is regrettable that classifications are presented as answers rather than as an identifying tool prior to explaining the diversity of existence they catalogue. Second, such an approach does not explain how groups come into existence or the motivation and rationale of individuals, bearing the same weaknesses as the "abstract self" of rational choice (utilitarian) theory:

> Theories of decision making . . . assume the pre-existence of preferences as providing a motivation for policy actors to select a particular course of action. They insist that the process of decision making can be understood by looking at actors' interests as prior attitudes to behaviour. . . . The fundamental deficiency of this model lies in the fact that it fails to concern itself with the origins of interest. It treats the interests adopted by policy actors as self-evident, ignoring the question as to how the alignment of particular interests and actors is actually determined. Politics of interest models of decision making cannot handle the question: "how do policy actors who behave in their own best interest come to know where that interest lies?"[39]

These rural actor classifications presuppose interests rather than how attitudes and perceptions change over time. Their explanatory power is hollow in that actions are taken as being motivated by interests but the perceptions underlying "interests" are ignored. Finally with the demise of landowner hegemony it does not show how each of the groups and individuals appeal to one another and interact. If no individual grouping can see all of its demands met in the fragmented environment without interacting to others, the question becomes how do they appeal to others?

Little research has been shed on this dimension, a notable exception being the work of Gordon Clark and his coresearchers with their studies of tourism and leisure conflicts in the countryside.[40] However, as outlined above, the reordering of rural communities goes well beyond the sphere of leisure and tourism so that the aim must be a holistic approach including these possible land uses with other economic and social contestations of space. Yet, Clark's work does give an indication of the different and competing conceptions of rurality that have emerged with growing accessibility, affluence, new technology, environmentalism, and the persistent desire of middle classes for a "rural way of life":

(1) a place of production (a factory floor) or a place of enjoyment (a playground);
(2) as private property or a communal resource;
(3) as a man-made environment or as a wild and natural place;
(4) as a place for people or somewhere to let off steam;
(5) as a managed or ordered place or as a place of freedom;
(6) as somewhere to be protected ("stewardship") because it is intrinsically valuable (e.g. landscapes) or as simply more land and resources to be used profitably ("exploitation");
(7) as a place where the normal rules apply (e.g., capitalism) or as *sui genris* where there are special rules;
(8) as a modern place or as somewhere old fashioned with old values;
(9) as a place to go or as a place to escape to.[41]

While it is not acknowledged by the author, each of these represents competing notions of the *telos* (end point) for what rural space should be like. Different notions of the end point thus imply a series of consequent, but differing, policy prescriptions. These competing notions of the *telos* are behind the fact that "dialogue, not to say controversy, is therefore inherent and inevitable in leisure uses in the countryside."[42] Central to our task, is thus the need to map the links between fragmentation and the way in which competing notions of the *telos* have emerged in their contemporary form.

At this point it is useful to consider what David Marquand, developing the analysis of Lindblom, outlines as the three modes of political change: command, exchange and preceptoral.[43]

Command Mode: Change is ordered from the top-down, actors operate as they are told to.
Exchange Mode: Change through the exchange relationships of individuals, change because it is made worth their while.
Preceptoral Mode: Change from persuasion, discussion, indoctrination and conversion—in short from learning. Change because the way in which individuals perceive and understand a situation has altered.[44]

Utilitarian thinking and rational choice has concentrated on the exchange mode for understanding politics, and all forms of the reductionist paradigm concentrate principally on the first two conceptions of politics. Under such a system of thinking, as Marquand argues:

> Hobbes, in some ways the greatest of all the thinkers in the reductionist tradition, painted a marvellously coherent, if chilling, picture of a society operating by the command mode. The seventeenth-century English Whigs, and their eighteenth-century American intellectual descendants, drove out their respective rulers in the name of the exchange mode. Both modes are, of course, omnipresent; it is hard to conceive of a society in which neither played a central part. Change can come by either of them, and frequently does. Almost by definition, however, neither can generate profound cultural changes—changes of value, belief and assumption.[45]

Crucially for the argument one develops here, it is possible to see how traditional agricultural interests have historically operated through the command and exchange modes via their domination of the rural economy, employment opportunities, and, in consequence, derived local political power. However, with the dilution of agricultural interests and the growing diversity of new actors, stemming from the reordering processes discussed above, the preceptoral mode has become far more important. If no single actor, or set of actors, constitutes a majority, the way in which these minority actors interact and appeal to each other becomes a vital area of study. To meet individual objectives actors must appeal to others, as no single set of actors can operate solely through the command and exchange levers. This is not to say these modes are unimportant or that relationships between actors are somehow now symmetrical, but rather, in any conceptualization of rural politics and studies of the interaction, relevant actors must consider how the various fragmented elements appeal to disinterested others and how in meeting their own objectives the actors make them seem in line with the overall aims of the whole community. The preceptoral level has become more important in this fragmented environment. Actors have to invoke notions of the "good rural society" and need to gain support from others for their view to meet their objectives.

Individualism and Communitarianism: Social Aims and Rural Communities

A broad and inclusive dichotomy can be drawn between individualist and communitarian theories of justice.[46] The former approaches contemporary normative political problems by assigning primary significance to the interests and values of individuals as distinct from the latter which views communities as the prime source of value. Each approach includes a myriad of different forms but, at the risk of oversimplifying, the overall debate encompasses two key controversies—a methodological and a normative discussion (see figure 9.1).* The methodological element of indi-

	Individualism	Communitariansism
Methodological	(I) The Abstract Self Atomism	(II) The Constructed Self
Normative	(III) Utilitarian Moral Cosmopolitanism Libertarianism	(IV) 'Moral Communities'

Fig. 9.1. Individualism and Communitarianism—A Framework for Analysis

vidualism has tended to rest on one or both of the following stances: (a) *the abstract self*, a way of conceiving individuals in which the features that "determine the ends which social arrangements are held (actually or ideally) to fulfill, whether these features are called instincts, faculties, needs, desires or rights, are assured as given, independently of a social context,"[47] and (b) *atomism*, which asserts that social phenomena can only be understood with recourse solely to individual actions (see cell I of figure 9.1).

These stances, which form the grounding for utilitarianism, have long been challenged (by Marx, Durkhein, and Veblen, amongst others) who claim that the individualist image of the self is ontologically false. To use the neo-Hegelian term of Chris Brown, the self is a "constituted construct"[48] (cell II, figure 9.1). The self, and thus individuals, are constituted by the community of which they are part. The community is thus more than an aggregate of individuals and has its own intrinsic value.[49] The implication for the normative debate is that the conceptions of justice people hold will derive from the communities to which they belong (cell IV). This leads many communitarians, but certainly not all, to conclude that global or absolute rules are therefore impossible. The community is itself a common good rather than the mere environment for individual interaction.

Not all those who aspire to global forms of justice (moral cosmopolitanism) hold individualist methodological views (cell III). Rawls, for example, in his early exposition of "justice as fairness" argues that, even if the individualistic case rests on false metaphysical premises, this provides no reason for rejecting the political solution of liberal neutrality.[50] Montefiore similarly argues that the answers to moral dilemmas can only come from a measure of detachment from one's own society—the community provides no basis for selecting between normative arguments.[51] Others continue to defend the methodological basis of individualism—restating the "primary rights" of individual autonomy, privacy and the dignity of man.[52] This debate, between individualists and communitarians as to the nature of ideal rules of, and rights within, societies has been reflected in notions of the ideal rural society.

The Rural Idyll

The rural idyll can be defined as the "collective images of what rural living should be."[53] The rural idyll embracing environmental and social integration has deep roots within British culture with the village seen as the ideal entity for fostering *"community* development":[54]

> Few people who live and work in the village have any doubt that social integration is worth fostering. In spite of the limits imposed on behaviour, the traditionalism, narrow-mindedness and gossip, village life offers something of high value—a sense of belonging to a face-to-face group.[55]

Victor Bonham-Carter concludes his book with the appeal, "Our aim must be an integrated rural community."[56] Richard Wollheim talked of an "English dream" of a collection of an unalienated folk society, bound together by tradition and by stable, local ties, symbolized by the village.[57] The perceived failure of many state-driven development projects within rural areas has increased interest in "bottom up" community-led development schemes for dealing with an array of local concerns from crime prevention to economic regeneration. More radical groups have taken this notion further, advocating small-scale participatory democracy as a basis for a new "environmental communitarianism," or bioregionalism.[58]

For the most part these conceptualizations of ideal rural communities have tended toward Romantic, post-Enlightenment notions of the ideal society. This is apparent by considering the nature of main rural movements in the U.K. and the date of their formation: the Commons Preservation Society (now the Open Spaces Society) was set up in 1865, the National Trust for Places of Historic Interest and Natural Beauty in 1894, the Society for the Promotion of Nature Reserves (now the Royal Society for Nature Conservation) began life in 1912, the Council for the Preservation of Rural England (CPRE) was established in 1926, and the Ramblers' Association started nine years later.[59] All these societies conform to the Romantic idyll with preservation as their main founding goal. Romanticism coupled with the Victorian religious revival created an environment in which Man was viewed as one of God's creatures, with the natural world given for mankind's use. By preserving natural environments Man was upholding the word of the Lord and rural communities were thus deemed ideal societies. However, while Romantic preservationist notions have dominated, other communitarian strains should not be ignored: the industrial village movement (associated with Rowntree), Owenite communities, bioregionalist groupings, or neo-Rousseauian participatory decision making of small "states."

Yet the rural idyll also embraces the heart of *individualism*—with rural living a "sense of achievement"; counterurbanization, to use Berlan-Darque and Collomb's term is very much about the achievement of "private ambitions."[60] For the new industrial middle classes of the Victorian period, rural landholding was especially appealing. The extent of this appeal and its implications for Britain's economic fortunes have been hotly contested, but its existence in some form cannot be questioned:[61]

> What started as a strongly moralistic critique of industrialism, of accumulation, and of the owners and managers of manufacturing industry had, by the end of the century, become a full-blown idealisation of the countryside, a ruralism which appeared to grow in inverse proportion to the importance of the rural economy itself . . . And F. M. L. Thompson (1990), in open dispute with Rubinstein, concludes that roughly sixty percent of business magnates worth more than half a million pounds at the time of death (between 1873–1875) acquired landed estates.
>
> The teachings of the public schools (Berghoff, 1990) and the Church of England, and above all the lure of the country house and the landed estate, meant that "at the moment of its triumph, the entrepreneurial class turned its energies to reshaping itself in the image of the class it was supplanting."[62]

More recently, the economic individualism of the 1980s had a clear rural focus with country houses and second homes, primary status symbols for those with high discretionary income. As Henry Porter comments on the gardening boom of the 1980s:

> . . . gardens laid out during the property boom of the Eighties, the suburban hybrid of two distinct traditions, one aristocratic, the other pastoral. In them the identifiable strands of English culture, a reverence for the past, together with a longing for the rural idyll and a fierce protection of private property.[63]

This was merely a new form of Conservatism, the roots of which are rural. Yet other individualistic dimensions of the rural idyll also exist that do not fall into the economic individualist camp—environmental cosmopolitanism, Kantian naturalism, and religious individualism.

The rural idyll thus embraces both communitarian and individualist values that are mutually exclusive. It does not in itself provide a *telos* for settling disputes between competing plans in rural localities. Crucially then, for all the long tradition of "rural worship," there has never been a single, clear, and consistent notion of what the countryside should be. The notions of what the countryside should be—rooted in both academic and lay discourse and expressed in both individualist and communitarian terms—reflect the deep divide in normative thinking rather than a consistent solution to the dilemma. In consequence there has never been a consistent benchmark by which competing notions of the good could be judged.

Historically in local rural politics, this divide between communitarianism and individualism has had less practical importance, as decision-making processes were dominated by the interests of agriculture with the notions that farmers believed to be the ideal rural society predominating. However, as with national decision making, agricultural interests have been squeezed out of their traditional strongholds in shire counties. Until the 1974 local county reorganization, farmers and landed interests were grossly overrepresented as councillors in rural areas (as indicated by the 1977 findings of the Robinson Committee) and were especially overrepresented as council leaders.[64] Up to 1945 most were "Independents," with, as Elcock writes, "council leaders . . . [being] major landowners their offices would almost be inherited with major estates."[65] As Lee found for Cheshire, authority tended to be vested in these traditional leaders by councils that were dominated by Independents with councillors exhibiting "traditional forms of deference."[66]

These agricultural groups watched their absolute power base dilute in the 1970s, as the 1974 merging of town and county (under the County Council reorganization) saw a displacement by town-based or newcomer Conservative leaders. These new leaders were predominantly drawn from shopkeepers, professionals, and local business owners and presented a more competitive approach to local politics. These new leaders of the 1970s and early 1980s prompted the subsuming of vast swaths of Independent councillors into the Conservative Party fold by issuing the ultimatum of "accept the Conservative whip or face a Conservative opponent at the next election."[67] However it was during the 1980s and early 1990s that the major shift occurred via the changing socioeconomic structure of the Conservative Party and its decline in local politics. At the time of the 1974 local government reorganization, Conservative councillors held over 40 percent of the county council seats (with still a large residual of like-minded Independents) and in 1977 controlled outright thirty-four of the forty-seven county councils in England and Wales.[68] By June 1995 Conservatives controlled outright only one county (Buckinghamshire) and eight local councils.[69] The dominant role of Conservatives in the local politics of rural and suburban Britain has now evaporated with the political arena populated by a diverse range of actors drawn from a multitude of backgrounds. The "farming idyll" no longer can be presumed to prevail in local decision making, with the scope and potential impact of alternative notions of the good society consequently enlarged.

Individuals and Communities in Fragmented Localities

While the communitarian tradition has long roots in academic and lay discourse, two central issues face its rural crystallization—the question of boundaries and, derived from this, the distinction between insiders and outsiders. When studying the social boundaries of rural communities, the interactionist approach outlined by Sarah Harper is of interest.[70] She argues that some individuals confine all aspects of their lives within one settlement, i.e., the physical (use of settlement—work, residence, and recreation), social (networks of kinship and friendship), and symbolic (past and present experiences, and thus images, of place). These individuals are termed "centered residents." In contrast, those who conduct little of their lives within one settlement may be described as "partially centered" or "noncentered" residents. In terms of settlements, one can thus distinguish between localities where the majority of the inhabitants "center" their lives on that settlement, which is thus a "centered" settlement, and other settlements that contain mainly "partially" or "noncentered" residents (in "partially centered" or "noncentered" settlements). Harper concludes by arguing that only those residents who center their lives on the settlement can claim membership of that settlement, thereby proffering that a real rural person is a centered person, residing in a centered settlement.

While Harper does not herself make the connection, the increasing diversity and mobility of residents, combined with contemporary production systems, means that very few residents fit into the centered groups. The centered community has become more problematic. The influx of counterurbanites and globalization of production systems means that very few will live, work, and socialize in the same community. Individuals belong to more than one community—a person may work in a city office (with its own community), join a local action group, go sailing with old university friends, and have family ties with people who live in another region. The reordering of rural communities means that the density and intensity of in-group relations within specific, spatially bounded microcommunities has declined.[71] In short, the totally "centered community," described by Harper, is dead.

The ideal of spatially bounded rural communities is no longer realistic under the contemporary political-economic framework. This is the loss behind what Harold Newby calls "rural retrospective regret."[72] Taking this notion of rural retrospective regret further, it can be seen as a community ideal that is desired yet no longer obtainable. However, the desires of the new counterurbanites, with their multicommunity belonging and maintenance of external contacts long after moving, are not supportive of this ideal. While the (mythical) golden age of "everybody knowing each other" and a thriving community centered on the village may be missed, individuals cannot and/or will not (and should they have to?) give up their multicommunity status. At the heart of the "village community" is thus a paradox—it is dead but mourned, desired but rejected.

Yet while a centered community is unobtainable, this does not mean that a spatially bounded, social sphere does not exist. The competing and often mutually exclusive possible uses of the countryside means that the private sphere inevitably enters the public realm. Externalities are pervasive. Individual and community notions of the good inevitably will clash—a result instantly verified by the "drawbridge mentality" of many counterurbanites. While defending the individual right of free movement of people (in terms of their own entry into the village), they subsequently wish to preclude this right to others, in the form of more development, in the name of preserving the local community. Moreover, with the growing diversity of actors, conflicts emerge between industries and services tailored to meeting the demands of these new rural participants. As Marsden et al. note:

> . . . the corporate house-building industry seeks out available land for "exclusive" housing, and the leisure industry purchases land for golf and other sporting complexes. Such interests are guided in their investment decisions both by the increasing demand for "rural" pursuits, experiences and values, and by the historical attractiveness and "authenticity" of particular rural places. They transform rural areas as they simultaneously seek to reproduce their scarcity, heritage, vagueness and desirability as refugees from a chaotic, urban world.[73]

The process of migrating to the countryside consumes the finite commodity of "rurality" in that it contributes to the development urbanities seek to escape from. It is recommended that future research concentrate on the implications of the paradox of the rural idyll—that it is at once desired but rejected. First, it is clear that individual preferences cannot be pure signals for community and social goals. While many wish to see integrated, small, and centered communities, they would not wish to be subject to such boundaries themselves and give up their multicommunity status. These individual preferences to be the exception to the rule cannot form the basis for just community decision making. If one takes David Miller's definition of social justice that the "just state of affairs is that in which each individual has exactly those benefits and burdens which are due to him by virtue of his personal characteristics and circumstances," it is clear that such exceptionalism is incompatible.[74] Second, it is apparent that multicommunity belonging and externalities mean that spatially bounded localities are not an entirely adequate unit for reconciling preferences. If the traditional image of the ideal rural community—the village—is not a discrete unit for reconciling preferences, the question immediately becomes what obligations does the community have to outsiders and how can political processes be structured to deal with activities that occur on several spatial levels? This links how notions of the good society always have a spatial dimension: if the decisions taken in one community affect others, political processes have to be designed to deal with the consequences of these overlaps. The reordering of rural localities with the influx of more mobile migrants and the declining socioeconomic and political importance of farmers has brought these questions of the good rural society to the fore. Within the U.K. there is not an adequate political framework for reconciling these preferences, and, given the contradictions inherent in many preferences, individual self-interests cannot perfectly signal collective goals.

Conclusions

There has been a greater and growing diversity within and between rural areas with a consequent increase in the number of potential conflicts as to the use of space. Individual and community-based theories have inevitably come into conflict. We are thus left with a fragmented social arena of competing possible uses of space but without a bounded community in which problems can be solved through consensus. The diversity of economic existence within given communities means that such a consensus cannot exist, and without clear and absolute boundaries any ethically defensible decision-making process must also take into account the rights of those outside the boundary. The rural idyll often appealed to, does not in itself provide this appropriate framework for evaluating between competing notions of justice. There has never been a clear and consistent historical tradition of what the countryside should be, for whom, and what should be the rights of those living within and outside it.

The breakdown of agricultural hegemony has heightened the importance of the preceptoral mode as individuals and groups have to attract support from, and convince or reassure, others. With the growing heterogeneity of rural actors in terms of both economic and sociodemographic status, their competing claims and perspectives represent an ever-widening spectrum. No clear rural political system or philosophical framework is in place to adjudicate between these claims.

Notes

1. M. Shoard, *This Land is Our Land: The Struggle for Britain's Countryside* (London: Paladin, 1987).

2. S. Harper, "People Moving to the Countryside: Case Studies of Decision Making," in *People in the Countryside: Studies of Social Change in Rural Britain* ed. T. Champion and C. Watkins (London: Paul Chapman, 1991).

3. P. J. Cloke and J. K. Little, *The Rural State? Limits to Planning in Rural Society* (Oxford: Clarendon Press, 1990).

4. D. Phillips and A. Williams, *Rural Britain: A Social Geography* (Oxford: Basil Blackwell, 1984).

5. P. Robert, P. and W. G. Randolph, "Beyond Decentralization: the Evolution of Population Distribution in England and Wales 1961–81," *Geoforum* 14, no.1 (1983): 75–102, 87.

6. Department of the Environment/Welsh Office, "Rural Enterprise and Development," *Planning Policy Guidance Note 7*, January 1988.

7. M. Sant and P. Simons, "The Conceptual Basis of Counterurbanisation: Critique and Development," *Australian Geographical Studies* 31, no.2 (1993): 113–26, 124.

8. R. Pahl, "Urbs in Rure: The Metropolitan Fringe in Hertfordshire," *London School of Economics Geographical Papers* 2 (1965).

9. H. Buller and P. Lowe, "Rural Development in Post-War Britain and France," in *Rural Studies in Britain and France*, ed. P. Lowe and M. Bodiguel (London: Belhaven Press, 1990), 21–36.

10. P. Boyle, "Changing Faces: Contrasting Residents and Newcomers in Remote Rural Areas." (Paper presented to the *IBG Migration Issues in Rural Areas Conference*, Swansea, England 1995).

11. Ibid., 12.

12. N. Bolton and B. Chalkley, "The Rural Population Turn-around: A Case Study of North Devon," *Journal of Rural Studies* 6, no.1 (1990): 29–43.

13. H. F. Marks, "Statistical Survey." in *A Hundred Years of British Farming*, ed. A.K. Britton (London: Taylor and Francis, 1989), 101.

14. G. Clark, "People Working in Farming: The Changing Nature of Farmwork," in *People in the Countryside: Studies of Social Change in Rural Britain*, ed. T. Champion and C. Watkins (London: Paul Chapman, 1991), 67–83.

15. National Audit Office, *Report by the Comptroller and Auditor General: Achievements and Costs of the Common Agricultural Policy in the United Kingdom* (London: HMSO, 1985): 3.

16. Ibid.

17. Department of the Environment/Welsh Office, "Rural Enterprise and Development": 4.

18. S. Whatmore, R. Munton, and T. Marsden, "The Rural Restructuring Pro-

cess: Emerging Divisions of Agricultural Property Rights," *Regional Studies* 24, no.3, (1990): 235–45.

19. MAFF, *Agricultural Census 1993* (London: HMSO, 1994).

20. G. Day and D. Thomas, "Rural Needs and Strategic Response: The Case of Rural Wales," *Local Economy* 6, no.1 (May 1991): 35–47; J. H. Gilligan, "The Rural Labour Process: A Case Study of a Cornish Town," in *Locality and Rurality: Economy and Society in Rural Regions*, ed. T. Bradley and P. Lowe (Norwich, England: Geo Books, 1984), 91–112.

21. G. Cox, P. Lowe, and M. Winter, "Agriculture and Conservation in Britain: A Policy Community Under Siege," in *Agriculture: People and Policies*, ed. G. Cox, P. Lowe, and M. Winter (London: Allen & Unwin, 1986), 181–215, 185.

22. N. Walford, "Agriculture in the Context of the Restructuring of Rural Employment," in *Contemporary Rural Systems in Transition: Economy and Society*, ed. I. Bowler, C. R. Bryant, and D. Nellis (Wallingford, Oxfordshire: CAB International, 1992), 187–200.

23. A. G. Jordan W. A. Maloney, and A. M. McLaughlin, "Characterizing Agricultural Policy-making," *Public Administration* 72, no. 4 (1994): 505–26, 506.

24. I. Hodge and S. Monk, "Manufacturing Employment Change within Rural Areas," *Journal of Rural Studies* 3, no.1 (1987): 65–69.

25. D. Keeble, S. Walker, and M. Robson, *New Firm Formation and Small Business Growth: Spatial and Temporal Variations and Determinants in the United Kingdom* (London: Employment Department, 1993), Research Series No.15: 52.

26. S. Grimes, "Indigenous Entrepreneurship in a Branch Plant Economy: The Case of Ireland," *Regional Studies* 27, no.4 (1993): 484–89.

27. A. Errington and R. Tranter, *Getting Out of Farming? Part Two: The Farmers* (Reading, England: Farm Management Unit, University of Reading, 1991): study no. 27.

28. P. M. Townroe and K. Mallalieu, "Founding a New Business in the Countryside" in *Small Firms in Urban and Rural Locations* ed. J. Curran and D. J. Storey (London: Routledge, 1993), 17–53.

29. D. Keeble, P. Tyler, G. Broom and J. Lewis, *Business Success in the Countryside: The Performance of Rural Enterprise* (London: HMSO, PA Cambridge Economic Consultants Ltd. for the Department of the Environment, 1992).

30. Central Statistical Office, *Regional Trends 27*, (London: HMSO, 1992), 76.

31. Ibid., 78.

32. R. Frankenberg, *Communities in Britain: Social Life in Town and Country*, rev. ed. (Harmondsworth, England: Penguin, 1973).

33. K. M. Halfacree, "Locality and Social Representation: Space, Discourse and Alternative Definitions of Rural," *Journal of Rural Studies* 9, no.1 (1993): 23–27, 25.

34. M. Young, and P. Wilmott, *Family and Kinship in East London* (London: Routledge and Kegan Paul, 1957); J. Connell, *The End of Tradition: Country Life in Central Surrey* (London: Routledge and Kegan Paul, 1978); Pahl, "Urbs in Rure."

35. Connell, *The End of Tradition*; E. Radford, *The New Villagers: Urban Pressure on Rural Areas in Worcestershire* (London: Frank Cass, 1970); J. Strak, *Rural*

Pluriactivity in the UK, A Report to the Rural Employment Group of the NEDC Agriculture Ad Hoc Sector Group (London: NEDO, 1990); M. Strathern, *Kinship at the Core: An Anthropology of Elmdon in the Nineteen Sixties* (Cambridge: Cambridge University Press, 1981).

36. Strak, *Rural Pluriactivity in the UK*, 41.

37. P. Ambrose, "The Rural/Urban Fringe as Battleground," in T*he English Rural Community: Image and Analysis* ed. B. Short (Cambridge: Cambridge University Press, 1992), 175–94; Pahl, "The Social Objectives of Village Planning"; G. M. Robinson, *Conflict and Change in the Countryside: Rural Society, Economy and Planning in the Developed World* (London: Belhaven, 1990); M. Shucksmith, *Housebuilding in Britain's Countryside* (London: Routledge, 1990).

38. G. M. Robinson, *Conflict and Change in the Countryside*, 42.

39. M. Schwartz and M. Thompson, *Divided We Stand: Redefining Politics, Technology and Social Choice* (London: Harvester Wheatsheaf, 1990), 49.

40. G. Clark, "Tourism and Leisure Conflicts in the British Countryside: the Cultural Dimension," in *The Retreat: Rural Land Use and European Agriculture* ed. E. C. A. Bolsius, G. Clark and J. G. Groenendijk (Utrecht: Netherlands Geographical Series, no.172, 1993), 109–19.

41. Ibid., 113.

42. Ibid., 114.

43. D. Marquand, *The Unprincipled Society: New Demands and Old Politics* (London: Jonathan Cape, 1988); C. E. Lindblom, *Politics and Markets: The World's Political-Economic Systems* (New York: Basic Books, 1977).

44. Marquand, *The Unprincipled Society*, 229.

45. Ibid., 228.

46. C. Brown, *International Relations Theory: New Normative Approaches* (London: Harvester Wheatsheaf, 1992).

47. S. Lukes, *Individualism* (Oxford: Basil Blackwell, 1973).

48. C. Brown, "The Ethics of Political Restructuring in Europe—The Perspective of Constitutive Theory," in *Political Restructuring in Europe: Ethical Perspectives*, ed. C. Brown (London: Routledge, 1994), 163–86.

49. S. Avineri and A. de-Shalit, eds., *Communitarianism and Individualism* (Oxford: Oxford University Press, 1992), "Introduction."

50. J. Rawls, "The Priority of Right and Ideas of the Good," *Philosophy and Public Affairs* 17, no.4 (1988): 251–77.

51. A. Montefiore, *Neutrality and Impartiality: The University and Political Commitment* (Cambridge: Cambridge University Press, 1975), 66.

52. D. Gauthier, *Morals By Agreement* (Oxford: Oxford University Press, 1986).

53. H. Newby, *Green and Pleasant Land? Social Change in Rural England* (Hounslow, England: Hutchinson, 1985).

54. M. J. Wiener, *English Culture and the Decline of the Industrial Spirit 1850–1980*, (Harmondsworth, England: Penguin, 1986), 41–80.

55. R. Crichton, *Commuters' Village: A Study of Community and Commuting in a Berkshire Village* (Dawlish, England: David and Charles, 1964), 9.

56. V. Bonham-Carter, *The English Village* (Harmondsworth: Penguin, 1952).

57. Wiener, *English Culture*, 42.

58. K. Sale, *Dwellers in the land: A Bioregional Vision*, (San Francisco: Sierra Club, 1985).

59. P. Lowe, "The Rural Idyll Defended: From Preservation to Conservation," in *The Rural Idyll*, ed. G. E. Mingay (London: Routledge, 1989), 113–31.

60. M. Berlan-Darque and P. Collomb, "Rural Population—Rural Vitality," *Sociologia Ruralis* 31, no.4 (1991): 252–61.

61. W. D. Rubinstein, "Cutting Up Rich: A Reply to F. M. L. Thompson," *Economic History Review* 45, no.2 (1992): 350–61.

62. M. Winter, *Rural Politics: Policies for Agriculture, Forestry, and the Environment* (London: Routledge, 1996), 176–78.

63. H. Porter, "The English Plot," *Guardian Review* 23 June 1995, 2–4.

64. A. Norton, "The Evidence Considered," in *Political Leadership in Local Authorities* ed. G. W. Jones and A. Norton (Birmingham, England: INLOGOV, 1978), 206–32.

65. H. Elcock, *Local Government: Policy and Management in Local Authorities*, 3d ed. (London: Routledge, 1994), 81.

66. J. M. Lee, *Political Leaders and Public Persons* (Oxford: Clarendon Press, 1963).

67. Elcock, *Local Government*, 80.

68. C. Rallings and M. Thrasher, *Local Elections Handbook 1993: The 1993 County Council Election Results in England and Wales* (Plymouth, England: Local Government Chronicle Elections Center, University of Plymouth, 1993), v.

69. R. Munton, "Regulating Rural Change: Property Rights, Economy and Environment—A Case Study from Cumbria, U.K.," *Journal of Rural Studies* 11, no.3: 269–84, 274.

70. S. Harper, "Rural Reference Groups and Images of Place," in *Humanistic Approaches in Geography*, ed. D. Pocock (Durham, England: University of Durham, Department of Geography, 1987), Occasional Paper no. 23; S. Harper, "The British Rural Community: an Overview of Perspectives," *Journal of Rural Studies* 5, no.2 (1989): 161–84.

71. C. Calhoun, "Indirect Relationships and Imagined Communities; Large Scale Social Integration and the Transformation of Everyday Life," in *Social Theory for a Changing Society*, ed. P. Bourdieu and J. S. Coleman (Boulder, Colo.: Westview, 1991).

72. H. Newby, "Revitalising the Countryside: the Opportunities and Pitfalls of Counterurbanisation," *Royal Society of Arts Journal* (August 1990): 630–36.

73. T. Marsden, J. Murdoch, P. Lowe, R. Munton, and A. Flynn, *Constructing the Countryside* (London: UCL Press, 1993), 10

74. D. Miller, *Social Justice* (Oxford: Clarendon, 1989), 20.

The Rights of Rights of Way

Hugh Mason

One of the minor features of the U.K. government's drive to divest itself of state involvement in the economy has been the selling of state (Forestry Commission) forests to the private sector. This has resulted in a sharp, often contested, change in the ease of public access. The Forestry Commission generally operated a freedom to roam policy whereby public access was not constrained except where young trees had been planted. The new private landowners have often rescinded the freedom to roam and invested heavily in barbed wire and fences. However in general they have been punctilious in respecting rights of way through their forests and not sought to prevent public access on foot or, in the case of bridle ways, on horse.

Both of these levels of access have been defended as being public rights but clearly they are different in that one can be arbitrarily but openly removed while the other commands, however grudgingly on the part of the new owners, a degree of respect. If a right of way is in some sense a right, in what sense, and how is it different from a freedom to roam. Both raise the basic questions of what constitutes a right, under what circumstances can any person obtain access to the landscape, and under what circumstances can access be denied.

Although private rights of way, or easements, where the right to access is vested in a particular individual or organization are common throughout the United Kingdom, in England the term is most generally used to refer to public rights of way where, although access may be limited to certain types of locomotion, the access is available to everyone.[1] A carriageway allows the right of passage to all wheeled traffic, and unless specifically excluded, the rights of a footpath or bridle way. A bridle way allows the passage, by riding or leading, of animals and, in modern times under an act of 1968, the riding of pedal bicycles. A footpath is, as its name implies, confined to pedestrian traffic, although rarely has this in practice excluded perambulators and wheelchairs, although in most cases the footpath would not be surfaced to permit such traffic and many are controlled by stiles and gates. These definitions lead to challenges from time to time, particularly from the riders of all-terrain bicycles, who often face the wrath of walkers when riding on a footpath[2].

The terms footpath and bridle way may raise the picture in the mind of a well defined track in a rural landscape, but this is not necessarily the case. Often, and indeed increasingly, the right of way may be an imaginary line from a gate to a stile across a field that may be pasture but that could equally likely be for growing wheat or cabbages. If a farmer expects regular use of the path, s/he may roll a path across the field but equally the walker may follow a nonexistent line across a muddy ploughed field or through growing crops. A right of way may traverse parkland to within a few feet of the windows of a stately home. Because many historic pathways have, during the course of time, become roads for general use, the right of way may just as likely now be a multi-lane highway in a heavily built-up area. The term "right of way" refers to a wide variety of pathways, all of which, however, allow free access without the payment of any toll for their use.

Although the public may look to the law to protect existing rights of way, the origin of the right does not lie in a legal provision. The origin of most footpaths and bridle ways lies in functional paths between villages or between settlements and churches, mills, or outlying houses. This has a number of consequences. Few rights of way lead to headlands, natural curiosities or other such potential tourist attractions, as these places were of little consequence historically. Indeed the path to Lands End, the most south-westerly point of the English mainland, from the main road is a private path to which the public has no right of access. Likewise there are few rights of way across moorland or highlands as these areas historically had low populations, few villages, very extensive land use, and thus little need for recognized pathways for man or beast. The demand in the twentieth century for long-distance footpaths and cycle ways for purely recreational purposes has been hampered by the lack of rights of way in remoter areas. In many cases these new routes have been cobbled together by connecting rights of way through the lengthy process of negotiation and persuasion of landowners to designate a path across their land or by the use of such features as abandoned railway tracks. In either case, the pathway may not be designated as a right of way and may in theory or in practice be closed for short periods of time. Although there is thus a generally used path, there exists no right to that path.

A further consequence is that the density of rights of way is far from uniform across England. The county with the densest pattern has more than four times the density of that with the sparsest. In addition to the determinants of the physical landscape, this variation is accounted for by a number of historical features, including the pattern of settlement, the pattern of landholding imposed when the medieval open field system was enclosed, and the local concern to preserve rights of way.

All perceived rights require general recognition and acceptance if they are to be effective. Amongst more fundamental human rights, the individual's right to suffer arrest and imprisonment only in consequence of his own actions requires a recognition and acceptance of the individual's worth as distinct from and greater than his position as a member of a clan, tribe, or family. The right of passage along a pathway similarly requires an acceptance and general recognition of that right as being greater than the landowner's right to enjoy the unhampered use of her or his own land. Certainly since the intensification of English farming in the latter part of the eighteenth century and probably before, there has been a conflict between these conflicting rights. As early as 1826 a society was in existence for the preservation of ancient footpaths to counteract the blocking of long-standing rights of way. Although the Prescription Act of 1832 allowed for the legal recognition of private rights of way, throughout the nineteenth century, despite the fact that public rights of way were generally acknowledged to exist, they enjoyed no legal protection.[3]

It was another hundred years before the Rights of Way Act of 1932 provided a means by which a public right of way might be established in law. The National Parks Act of 1949 simplified the procedure so that any path that had been used unchallenged for twenty years and where there was no evidence to show that the landowner had wished to prevent the path being dedicated to the public could be established as a right of way. Following this act an attempt was made to map all rights of way in England and Wales: blank maps were sent to every parish council and the results collated on a county level. This exercise did not prove definitive, however, because some parishes were more diligent than others, and in some cases it was left to the chairman, or even a large landowner, to complete the draft. Although in many cases paths were added after the first publication of the map, once a path had been omitted the process of getting it reinstated was both lengthy and complex.[4]

The act also allowed for the creation of rights of way by legal process even where there was no tradition of an existing right or even the consent of the landowner. A path through Wychwood Forest in Oxfordshire that had previously always been closed for a day a year to prevent its becoming a public right of way was created such in 1986 by a government order. Moreover, under current legislation it is possible to have the route of a right of way diverted or even stopped up, but the request for this must be made to the magistrates' court by the local authority, and although the local authority may be persuaded to do this by the landowner, he or she has no ability to initiate proceeding on his or her own account.

Footpaths still exist, however, that are in general used almost continuously but that remain permissive routes rather than rights of way. For example, in the city of London there is a pathway through Lincoln's Inn that is much used every day of the year except Ascension Day, when it is closed specifically to prevent it from becoming a right of way.

Rights of way are thus routes that are valued and defended, often with great passion and emotion even in cases where convenient alternative routes exist. They are generally considered to be more than convenient legal creations: they are considered to be paths that may be used, for business or for recreation, without let or hindrance, by anyone at any time, and with the expectation that the paths will remain open in perpetuity. Yet equally defended by some are the historic distinctions, and woe betide the cyclist on the footpath or the motorcyclist on a bridle way, for the rights, like many things that are considered sacrosanct, are quite specific.

The rights that are claimed and defended are therefore rights of public access through private property. Nothing in the concept of a right of way questions the right to private property, indeed the reverse is true—the right of passage is exceptional and must be defined, while the right to property is assumed and the respect of that property is considered normal. This is true even where the ownership of the land is in common, or where the state is the owner. Common ownership and state ownership are for this purpose equivalent to the private ownership of land, as all have the right in law generally to prevent public access to their territory.

While the right of way shares many characteristics with other rights, it does not fit neatly into the common classifications of rights. Most such categorized rights have their origin either in the existing legal framework or in a wider moral framework of basic human rights that should be, but rarely are, protected for every individual.[5] The right of way fits easily into neither category for no one would claim it as a basic human right and in its origins it is more than a legal product having been defended against the provision of English law throughout much of the nineteenth century and indeed having been pressed by quite deliberate breaking of the law upon occasions. Rather, the source of the right lies in its antiquity, the particular pathway having been used freely for such a length of time that a right to its continuance is generally regarded as right and proper. It could therefore be viewed as society's statement that a liberty that has existed for a very long time should ipso facto continue to exist as a customary right and not be subject to arbitrary removal. This principle would appear to have been recognized, although not created, by the legal provision that allows for the creation of a new right of way after twenty years unhindered use, provided that no intention on the part of the owner to prevent it can be demonstrated. This principle of antiquity and continuity producing a customary right is not unique. The right of anyone to speak their mind, however much it might incense the Crown, Parliament, or the legal profession, at London's Hyde Park Corner and a few other places in England has no legal standing but has existed for a sufficiently long time that it is generally considered to be a customary right.

The right of way, similarly, neither falls easily into the category of a negative, or protective, right that limits the treatment a government or ruler can mete out, nor is it a positive right equivalent to the right to particular welfare goods, although a case could be made for either. Indeed, in terms of how the right is regarded within society, it appears to have changed its status from being primarily a protective right, whereby a person could not be removed from a pathway by the agents of any landlord, to being in most cases a positive right to access a path through the countryside for recreation or merely for the convenience of not taking the long way round.

This then raises two questions. First, is a long precedence really sufficient to allow anything to continue to be regarded as a right? And, second, may a right change its function and purpose and still be regarded as the same right?

It is certainly the case that rights are claimed, as, for example in the United States, the right to carry arms beyond the limited circumstances defined in the Constitution, on the basis of long-established and generally supported custom. However, although the right may be so defended, it does not prevent its being challenged on the ground that it conflicts with other rights that should take precedence. In this case, the counterargument would surely be that, although the antiquity of the right is recognized, circumstances have changed now bringing that right into conflict with other rights that must be regarded as taking precedence, such as, the right to life and to domestic security. In the case of a right of way, the same principle would seem both to be appropriate and to apply in practice both in particular and in more general cases. A footpath or bridle way may be stopped or diverted if a need can be demonstrated that should take precedence over the right. Indeed, this has occurred generally where a whole tract of land has been taken for military purposes or commercial excavation.[6] Whether or not one regards the right in utilitarian terms, the argument for stopping or diverting is usually presented in that way: the needs of society as a whole takes precedence over the rights of that part of society that wishes for substantial or frivolous reasons to use the pathway or pathways. It would thus seem that long precedent is sufficient to support a claim to a particular right although it does not give that right an absolute status with respect to other rights.

If the right may be removed or modified in specific cases, it would also appear that the substance of the right may develop and change through time. Indeed, in this respect the right of way is no different from a number of other positive rights. The substance of the right to health care not merely differs through space from country to country but changes through time with different political philosophies and technological capabilities. This right has in many countries developed over the past decade to incorporate a corresponding duty on the individual to bear a part of the cost involved. There would therefore appear to be nothing abnormal in the substance of the right of way being modified from protecting a simple local convenience to a more general and recreational access without disturbing the existence of a right. By the same token there would appear to be no reason why the right of access to footpaths should not be extended to encompass access by people on all-terrain bicycles or even electric wheelchairs. The matter would not appear to be one of principle so much as one of common courtesy between users. Indeed, since the cycle ways that have been created in England and Wales under the Cycle Tracks Act of 1984 are also pedestrian rights of way, the development would appear to be in process toward three levels of rights of way: footpaths, cycle ways, and bridle ways.[7]

Because the concept of the right of way has therefore changed through time and is now established within a framework of legislation, questions are raised of what the relationship is between the law and the right and moreover whether, in this case at least, the concept of a right as distinct from and more than a legal provision now has any value at all. The earlier legislation, until and including the 1949 Countryside Act, had four primary purposes: (1) it recognized the existence of rights of way, (2) it made provision for the designation of rights of way where they could be shown already de facto to exist, (3) it provided for the specific detail of their existence to be recorded, and (4) it imposed a duty on landowners and local authorities to recognize and to a certain extent to maintain that way.[8] The later legislation extends beyond recognition and consolidation to enlarge the principle, for example to new forms of locomotion.

The law thus has progressed beyond merely protecting a right, but there remains a substantial reason for maintaining the idea of the right rather than limiting the concept to a mere legal provision. What legislation creates, legislation can remove. Although a consequence of legislation rather than a specific provision of legislation, the freedom to roam through forests could be extinguished with their transfer from the public to the private sector. Were the law arbitrarily to restrict access to public paths or to extend the rights of landowners to enable them to charge for the use of such paths running across their land, the legal provision for rights of way would effectively be extinguished. There is however, no doubt that a substantial section of society would object and seek to maintain what they considered, by virtue of its antiquity, to be their right. If, as would be likely, this involved mass trespass, they would undoubtedly claim a moral justification for breaking the law. Thus, although pathways are now the subject of considerable legislation, the idea of a right that is distinct from, and in a moral sense has precedence to, the legal provision is a realistic one. Nevertheless, it is open to question how far from its original conception a right can be developed by legislation and still remain a right that can be defended against legislative change. The right is thus a statement of widespread will to regard something as just and to preserve it against all attempts to remove it.

The concept of the right of way may be viewed as a particular response to a distinctive situation; the right balances a right to the exclusive use and protection of property that is particularly well developed. In other contexts where the right to property is different, the requirement to define rights of access will be different. Although the intensity of land use appears to be a major consideration as, except where wildlife is being protected or specifically reared, there is little reason to prevent access to uncultivated land, the traditions of the country and its implicit or explicit understanding of what constitutes trespass and what rights people have to move at will seem to be of greater importance.

In much of sub-Saharan Africa, traditionally the ownership as distinct from the use of land could not be held by an individual. In many parts of Tanzania the ownership lay with the community, both living and dead, and except for agricultural land, could be used by any member of the community without restriction.[9]. Similarly amongst the Shona people of Zimbabwe, land was regarded as holy and thus not ownable.[10] Consequently the matter of access was not an issue unless damage was caused.

Even where European settlement or plantation agriculture has not effected a change, the growth of population and change in culture during this century has usually produced changes of practice in the holding of property. Among the Iteso of Uganda, there were by midcentury already signs of a tendency to allow prominent persons to acquire more land than they needed for cultivation, and with increasing population it is possible that such land will be denied to others.[11] Although in the last half century there has been a change to the individual or family ownership of land, the pathways to fields or settlements appear to have been unaffected.[12] This may be related to the fact that fields are often divided by paths but rarely by fences and that the majority of people still live on the land that they farm. It is, however, clear that for whatever reason the development of property rights has not been at the expense of the right of passage.

In Europe also the right of property was not necessarily exercised in a way that restricted movement. The restriction of access to established rights of way in England was certainly a cause of some surprise to visitors from the Baltics in the 1930s, who were used to less restricted access.[13] In other parts of Europe, particularly where moves to agricultural land consolidation have reduced the need for minor pathways, governments have been active in promoting recreational routes, particularly long-distance paths such as the *routes de longue randonèe* in France.

The English right of way is thus a particular response to distinctive circumstances. Alternatives to the legal recognition of historic rights and the extension of the pattern of historic routes could exist. The belief of the walker, horse rider, or cyclist that they are exercising a right born of long precedent is, however, at least as sure a foundation for public access as is specific and detailed legislation. It may therefore be concluded that the English right of way is more than an antiquated form of words. It describes what society believes to be the proper status of particular pathways and any person's right, irrespective of the vagaries of changing laws, to access those pathways to whatever end they choose. This right owes its existence to long-standing exercise of the substance of that right, and as a corollary the right needs to be exercised if it is not to fall into disuse. It is also generally accepted that rights are not necessarily static in their specific provisions: new applications may be developed to meet changing circumstances. However, the right extends beyond legal provision; the law may protect and preserve the right but cannot extinguish it.

Notes

1. J. F. Garner, *Rights of Way and Access to the Countryside*, 4th ed. (London: Oyez Longman, 1982), 1–10.

2. "MTB, 94," *Mountain Bike* (January 1994). This is also available on the internet at: http://www.futurenet.co/Access.html

3. D. Chapman, *Walmsley's Rural Estate Management*, 5th ed. (London: Estates Gazette, 1969), 205.

4. P. Clayden and J. Trevelyan, *Rights of Way—A Guide to Law and Practice* (London: Commons, Open Spaces and Footpaths Preservation Society, 1983), 28–65.

5. A number of sources have influenced the development of this part of the paper, in particular: B. Almond, "Rights," in *A Companion to Ethics* ed. P. Singer (Oxford and Cambridge, Mass.: Basil Blackwell, 1991); L. C. Becker, "Individual Rights," in *And Justice for All*, ed. T. Regan and D. VanDeVeer (Totowa, England: Rowman and Littlefield, 1982); R. Frey, *Rights and Interests* (London: Oxford University Press, 1980); R. Frey, *Utility and Rights* (Oxford and Cambridge, Mass: Basil Blackwell, 1985); A. I. Melden, *Rights and Persons* (Oxford and Cambridge, Mass.: Basil Blackwell, 1977).

6. Clayden and Trevelyan, *Rights of Way*, 79–85.

7. Cycle Tracks Act, Eliz 2, 1984, c. 38.

8. National Parks and Access to the Countryside Act, 1949, 12, 13, and 14 Geo 6, c. 97.

9. C. K. Omari, "Traditional African Land Ethics," in J. R. Engel and J. G. Engel, *Ethics of Environment and Development* (London, Belhaven, 1990), 168.

10. S. Bakare, *My Right to Land* (Hrarae: Zimbabwe Council of Churches, 1993), 47.

11. J. C. D. Lawrence, *The Iteso* (London: Oxford University Press, 1957), 242.

12. Personal communication from the Hon Grace Akello M.P., Wera, Uganda.

13. Personal communication from the late Leopold Kivimagi, Tartu, Estonia.

The Mediation of the Public Sphere: Ideological Origins, Practical Possibilities

John Stevenson

The impoverishment and degradation of the public realm has been a prime topic of concern and alarm for social critics throughout the postwar era—certainly in the United States since the debates on mass society and its culture in the 1950s and such books as Galbraith's widely read *The Affluent Society*, with its focus upon the contradiction between private affluence and the starving of the public sector.[1] In recent years much of the discussion of this topic on a theoretical plane has revolved around the concept of the public sphere, as developed by philosopher and social theorist Jurgen Habermas. This concept—combining theoretical, historical and normative aspects—speaks to a number of contemporary concerns in social and cultural theory and morality. Yet it is possible that the concept's widespread appeal results from a finessing, rather than a resolution, of certain antinomies of power and morality, of history and ethical norms, and from a privileging of print-based media of social communication. I want to explore these issues in relation to some cultural-political history from the not-too-distant past, as well to the contemporary project of reconstituting the public sphere.

In Habermas's usage, the public sphere in modern society is the social domain that mediates between civil society (the economic realm of commodity exchange and social labor) and the state.[2] Civil society, in classic social theory, is the field of the private (private property, as well as exchanges, contracts, and associations between and among private persons), whereas the state institutionally embodies the public life of society; in Habermas's theory this mediating area, the public sphere, is characterized as a sphere of private persons come together to form a public. It thus appears as the realm that effects an intermediation between private and public in modern societies or whereby private persons become public.

The public sphere, then, as an aspect of modern societies, is an abstract social space. As an analytic category—a theoretical entity, really—it cannot of course be identified with any of the material ways that it may manifest itself in practice. Thus, although there is certainly a relation between them, the public sphere must be distinguished from physical public space.[3]

Arising in the European seventeenth and eighteenth centuries as a realm of participation based on private autonomy and associated with the rise of the press as well as with the opening up and more general consumption of literary and general artistic productions (and the consequent coming into being of literary and art criticism), the public sphere underwent a structural transformation in the late nineteenth and early twentieth centuries. The growth of the extent of state functions and the way in which private institutions (e.g., corporations) have assumed a semipublic character have increasingly squeezed and eaten up the public sphere from both sides. By the latter part of the twentieth century a change has transpired, according to Habermas, "from a culture-debating to a culture-consuming public." Publicity, which originally assured that rational-critical public debate would have its effect upon the actions of the state, now serves to manipulate the public; it is at most ". . .the court *before* whom public prestige can be displayed—rather than *in* which public critical debate is carried on." With the advent of consumerism and the welfare state, the public sphere becomes the domain of advertising, public relations, and the engineering of consent.[4]

As a category of social analysis, the public sphere is thus historical— relevant only to modern societies and undergoing historical change within them—and class-specific as well. It is, says Habermas, a category of bourgeois society, created by a rising bourgeoisie and applicable only or mainly, we may presume, in societies characterized by bourgeois hegemony.[5] On the other hand, the public sphere is also associated with the public use of reason, and this fact endows the category with a certain transhistorical and ideal quality. It is not *simply*, in Habermas's construal, an historically relative category or institution, nor simply one more weapon in the struggle of social classes; it now appears to embody an essential, lofty, and (socially, morally) valuable characteristic of human experience, and thus achieves a transcendent ideality. When it is characterized in terms of the use of reason, the public sphere becomes not simply a category of historical description, but also of normative judgment. Thus Habermas's tracing of its structural transformation over the course of the last century is not only the description of a social change, but also a narrative of decline and a social critique—a tocsin warning of the sickness and immanent death of an invaluable social mechanism.[6]

In adopting this point of view, it is easy to demonstrate that the public sphere has only undergone an accelerated decline in the thirty five years since Habermas's book first appeared, with a simultaneous degradation of cultural-intellectual colloquy, the professionalization and academicization of intellectual discourse(s), a greater use of manipulative rationality in the public sphere (advertising, including especially political advertising, is an obvious example, but a good case can be made for saying that it permeates journalism and other forms of public discourse as well) and, generally, the increased separation on all sides from the use of reason in a universalistic way. Our cultural condition seems to have become more and more privatized, fragmented, and restricted.

These sorts of laments are not uncommon these days, but, in the first place, they cry out for greater context. What is the benchmark for the statement that this sort of decline has occurred? Is a golden age of the public sphere presupposed? To approach these questions, I want to map out a state of things, in the American 1940s, in which the public sphere may well be said to have enjoyed a relative flourishing. I will show that there is a good *prima facie* case for attributing a fairly vibrant Habermasian public sphere to the United States in this period, compared to today. Looking a little deeper, though, we will find several ways in which this public sphere was crucially constricted and compromised; in certain aspects its pretensions were spurious.

But (in the second place) if this is the case, doesn't it raise questions about the concept of the public sphere itself, and its applicability? Indeed, as we shall see, these sorts of problems in the public sphere of the 1940s interrelate with questions raised by some commentators on Habermas and by contemporary work in media and cultural studies: questions of the relative unity or multiplicity of "the" public sphere, on the one hand, and considerations attendant upon the rise of new media, on the other. Examining the difficulties generated by these sorts of inquiries, we will see problems raised for the Habermasian narrative of decline, as well as for our own, at least in its present form. Finally, I will use this dialectic of chronicle and critique to explore briefly the question of what it might mean to constitute anew a viable public sphere in the contemporary world.

America in the Forties: The Public Sphere and Its Transformation

Crucial parts of the American 1940s have disappeared, both in fact and in public memory, virtually without a trace. The recent fiftieth-anniversary-inspired recollections of the war years, for example, have not only tended to highlight only certain aspects and topics, but to ignore entirely the socio-cultural events and dynamics of the years immediately preceding and following American participation in the war. I will focus first on these years (roughly 1939–47), before the structural changes wrought by the hot and cold wars had taken effect. During most of these years, intense discussion and debate about national and global issues was taking place in a variety of public fora: national and local newspapers, popular and "middlebrow" magazines as well as more "highbrow" journals, books, radio, the local and national meetings of organizations and associations, and large public gatherings called for this purpose. Before the war the focus was on U.S. entry into it; wartime discussion concerned the shape of the postwar world; and postwar debate centered around the ramifications of the newly released power of the atom. In reviewing these episodes, it is hard to avoid the impression that the debates took place within a general social arena—a public sphere—that has since been subverted and shattered. At the same time one can discern during this era the forces at work that led to its decline or destruction.

The debates of the prewar period, on U.S. participation in the European war, are undoubtedly the most familiar. Interventionists v. isolationists: the battle has been described in many histories, usually to the benefit of the former, viewed as forward-looking and progressive, ready to shoulder the "responsibilities of world power," in contrast to the backward-looking ostrich isolationists, who selfishly and unreasonably wished America to cut its ties and turn its back on the world. Regardless of the truth of this picture, it neglects the reasons and arguments put forward on either side. For this was, above all, a wide-ranging public *argument*, with large numbers of people drawn into an increasingly polarized debate in which partisans at both poles sought to persuade and convince the mass of people in between.

The primary argument deployed on the interventionists' side was to the effect that European fascism menaced the security of the United States. Many, such as Robert Sherwood, Dorothy Thompson, Clarence Streit and other leading anti-fascist polemicists, advocating active intervention or even a federal union of the democratic or the English-speaking nations, held that Western civilization itself was threatened. Even Roosevelt, usually far more circumspect, averred, in a speech of May 1941, for example, that the originally purely European war had developed, "as the Nazis always intended it should develop, into a war for world domination." Clearly, if these were the stakes, the United States was bound to become involved.[7]

Isolationists, on the other hand, denied that European conflicts posed a threat to U.S. security, or argued that if they did, the answer was a strong hemispheric defense, not active participation in events on the other side of the Atlantic. They further claimed that, far from serving to protect

America, entrance into the war would lead to this country's becoming structurally similar to the fascist foes it would fight. Socialist Norman Thomas, for example, argued that the historic racism and violence of America would make it impossible for America to resist the development of fascism in time of war, while from the other side of the fence, Republican Senator Arthur Vandenberg, at that time inclined toward isolationism and a strong opponent of U.S. entry into war, wrote a friend that

> we cannot regiment America (and how!) through another war and ever get individual liberty back and freedom of action back again. We shall be ourselves a *totalitarian state*, to all intents and purposes, within ten minutes after we enter this war as a protest *against* totalitarian states.[8]

Their thesis was that the necessities and exigencies of modern warfare would dictate the sort of state and social system that was seen as responsible for making Germany the efficient war machine that had, by this point, conquered or cowed most of Europe. Conservatives feared that war would make possible a transition of the New Deal government to the socialist dictatorship toward which they'd always thought it tended, while liberal and radical isolationists believed war would spell an end to domestic reforms.

A vast public debate, it's clear, took place during these years around the question of U.S. intervention in the developing world war, and, although—as in any dispute of this magnitude and passion—emotive and fallacious forms of argumentation were often enough deployed, in the main the appeal was to reason. This period saw, it would seem, an exemplary use of the public sphere.

The same could be said, perhaps to an even greater degree, of our second period, encompassing the war years. Sense of the momentous nature of the war was widespread, and there was vigorous discussion and debate as to the sort of world that should or would emerge from the epic struggle. Liberal prescriptions for domestic and international reform were rife and felt to be mandated by the war. At the very least, as James Reston put it, "the reason we went to war is that we were attacked; the purpose of the war is to make sure that we create a world order in which we can never be attacked again." From this minimalist position, though, Reston goes on to deduce the necessity of "national economic planning by the government" and "some kind of economic and political collaboration with other powerful, peace-loving nations"—all of this undergirt by a movement "to regenerate the spirit on which the nation was founded and through which it made itself great."[9]

Variations on these aspects—economic reform, international cooperation, and spiritual or moral renewal—were widely debated within the liberal climate of opinion that dominated among the "talking classes," with positions ranging from world federalism to regional associations that would cooperate globally, from governmental economic planning and control to the breakup of corporate power and the unleashing of small business entrepreneurship—all accompanied by visions of peace, abundance, and an equitable distribution of goods. The champion of worldwide social liberalism in the political arena was Vice President Henry Wallace, whose 1942 speech proclaiming this the Century of the Common Man was issued in pamphlet form, translated into many languages, extensively disseminated at home and abroad, and later expanded by Wallace into a book. Republican liberalism was represented by Roosevelt's 1940 opponent Wendell Wilkie, whose widely circulated book *One World* proclaimed the necessity of international cooperation: "America must choose one of three courses after this war: narrow nationalism, . . . international imperialism, . . . or the creation of a world in which there shall be an equality of opportunity for every race and every nation."[10]

Further to the left, American Communists, who may have attained their greatest influence during the Second World War, when they were able to "fuse American and Russian patriotism," also participated in the discussion and debate. Communist leader Earl Browder was particularly impressed by the Tehran Conference, the first meeting of Roosevelt, Churchill, and Stalin, at which the outlines of a common strategy for the remainder of the war and afterwards was sketched. Browder saw the agreements reached as a portent of postwar cooperation between socialism and liberal capitalism, and thus as (in his words) "the greatest and most important turning point in history."[11]

The sense of a turning point in history, of momentous dangers and possibilities, was widespread: "One of two alternatives will confront the postwar world," said Ruth Nanda Anshen in the volume of essays she edited, *Beyond Victory*:

> Either civilization will be precipitated into a rapid, metastatic decay, war following upon war, or will even be violently overthrown—or it will experience a decisive transformation and new birth. It will all depend upon whether men have moral purpose enough to increase their need for social cohesion and mutual obligation. . . .

Moralists, educators, economists, anthropologists, political figures, psychologists, social scientists, and journalists all discussed and debated proposals and prescriptions for the postwar world.[12]

The voices in this debate were not all from the left-liberal side of the spectrum. The intellectual basis of postwar conservativism was laid during the war by such works as Friedrich Hayek's *The Road to Serfdom*, which received widespread attention, and John Flynn's *As We Go Marching*—both of which held that the welfare state was the first step to the fascism that the U.S. was fighting in Europe. In the political arena, the conservative reaction against New Deal liberalism often found voice within Congress. The 1942 elections, indeed, had brought large defeats for liberals, and the Congress that convened in 1943, controlled by a conservative coalition of Republicans and southern Democrats, immediately began dismantling New Deal agencies and harassing the more liberal administration appointees, whom the *Wall Street Journal* accused of making "totalitarian" plans "to make permanent changes in American social and economic institutions."[13]

Finally, at war's end, both of these debates coalesced around the new catalyst of atomic weaponry and power. In response to the dawning of what was immediately dubbed "the atomic age," the advocacy of some form of world government became widespread, with partisans ranging from Walter Lippmann to E. B. White and Norman Cousins. Even the *Reader's Digest* published an article declaring that "world government has now become a hard-boiled, practical and urgent necessity." Images of utopian possibilities to be achieved through the peaceful use of atomic power were also rife, as were, on the other hand, apocalyptic visions of world's end by atomic fire—visions employed both by a burgeoning movement of scientists mobilized to public action by the atomic threat and by evangelical protestant preachers like the young Billy Graham. Themes similar or identical to those articulated during the war years, then, involving the need for international cooperation or spiritual and moral renewal, as well as a sense of momentous dangers and possibilities, are now proclaimed in terms of the atomic age.[14]

It is worth noting, however, that although these three episodes can all be described as uses of the public sphere to debate momentous issues facing the social whole, it can also be argued that in not one of these cases did the actual debate determine the subsequent public policy adopted. The outcomes were not the result of excellence of reasons or reasoning, nor (arguably) of "the people's will" in the sense of majorities coalescing, through public-sphere debates, around a given policy. To an ever-greater extent, during this period, the state entered as a participant in public-sphere debates—increasingly as a covert participant—working to shape a public-sphere mandate for actions whose genesis lay elsewhere.

In the prewar debate, for example, the principal interventionist organization, cumbersomely but shrewdly titled Committee to Defend America by Aiding the Allies, had been begun by William Allen White at the president's urging, and it crafted its message in close consultation with the White House. Before the U.S. entrance into the war, the administration had experimented with several small governmental public relations agencies, designed to "stimulate patriotism," disseminate helpful news, and assess the public mood. During the war not only was there the Office of War Information and official censorship agencies but, more important, there was through both formal and informal channels unprecedented cooperation among government officials, news organizations, film studios, other entertainment networks, and large corporations and their advertising agencies in the presentation of a unified and coherent view of the war and its meaning.[15]

After 1947, when the cold war began to set its seal on the times, public-sphere debate was severely restricted. The advent of security/loyalty tests and the repressive frenzy usually known as McCarthyism effectively choked off or marginalized the expression of opinions beyond a narrowly defined band of orthodoxy. It was, moreover, not only a question of repression. In many areas of American life, domestic as well as foreign, the newly formed Central Intelligence Agency (CIA) and other state intelligence agencies began to operate clandestinely, sponsoring organizations, publications and studies, or assuming the public guise of journalists, government officials, businessmen, scholars, or clergymen. To take just one example, a major postwar publicity project within the United States sponsored and financed by the CIA—the Crusade for Freedom—had a budget larger than the combined total of all money spent on the presidential election campaign of 1948. The budget itself vastly underrates the actual effect on public life, for much of it was used to garner free radio, television, newspaper, and magazine advertising. Even so, this figure alone establishes the CIA as "the largest single political advertiser on the American scene during the early 1950s, rivaled only by such commercial giants as General Motors and Proctor & Gamble in its domination of the airwaves." The result of this and numerous other covert state security and intelligence operations was to establish a very large and secret netherworld that came to dominate and swallow up more and more of the public sphere.[16]

Much of this clandestine activity was exposed beginning in the latter half of the 1960s, a time that also saw a new efflorescence of public debate. Since then—despite the exposures, debates, and bitter struggles against state repression during the sixties—the role of the state within the public sphere decreased only minimally while, on the other hand, undercover governmental activity has once again been on the rise. This time has also seen the increasing (and increasingly sophisticated) influence and power of corporations in both the media (through advertising) and the state (through extensive lobbying and the financing of political campaigns).

It may be argued that many other factors have gone into the decline of the public sphere (some of which will be discussed below): the rise of television, the conglomeration of the press, the retreat of intellectuals to the academy, and a fragmentation of *the public* into a host of interest groups. If I have emphasized the role of the state (and its interlock with increasingly multinational corporations), it is because this aspect is often downplayed or virtually overlooked: the extent to which the burgeoning powers of the national-security state have been instrumental in determining (or interacting heavily with) the form, extent, and content of public intellectual life.

The Public Sphere: Questions, Tensions and Contradictions

The decline of the American public sphere seems to present itself as an evident fact, and we have seen that an excellent *prima facie* case can be made for such a judgment, both over the last century and a half and over the past fifty years. Indeed the delineation of its sorry, tattered, and degraded state might be said to constitute a genre of contemporary American social criticism.[17] So strong and compelling are these aspects of contemporary social reality that they may blind us to the sorely problematic nature of the public sphere, even in its state of relative flourishing in the American 1940s. It is not simply that creeping behind-the-scenes state influence, as sketched above, lends a certain spurious quality to these momentous debates, at least as regards their effects upon state actions. It is also that the public sphere of the forties, even at its most legitimate, was false in its self-representation as a universally accessible field, for in actuality it was, de facto if not de jure, fenced in by restrictions of race, gender, and class that render problematic its own inherent ideals. Participants in the public debates of the decade were almost entirely white, male, and middle (or upper) class, and in this sense the public sphere of the time fell short of its own implicitly self-proclaimed identity as a realm of discussion to which all had access and in which ideas were judged according to criteria of reason and argumentative support. Although there were notable postwar breakthroughs by such writers as Richard Wright and Mary McCarthy, as well as by some from working-class backgrounds, the realm still remained overwhelmingly the preserve of men who had been more or less born to it.

The same could be said—and to an even greater extent—about the eighteenth century public sphere that Habermas delineates. As he makes clear, the status of the private man who might participate in the public sphere was that of the male head of a household made up of both property and family, who thus ". . . combined the role of owner of commodities with that of head of family, that of property owner with that of 'human being' per se."[18]

It is obvious that anyone today who would propose a repair or reconstitution of the public sphere would want to eliminate any such inegalitar-ian constraints. The matter, however, is not quite so simple, and upon reflection the restrictedness of the public sphere historically raises more questions for the project of rebuilding this realm than would at first ap-pear. A thesis raised by several recent commentators is that there have actually been, historically and today, not one but many public spheres: "[V]irtually contemporaneous with the bourgeois public there arose a host of competing counterpublics, including nationalist publics, popular peasant publics, elite women's publics, and working-class publics."[19] This fact, along with the observation that these counterpublics have been com-prised of groups excluded from the official public sphere, raises the ques-tion of whether a drive to rebuild a single hegemonic public sphere is the best way to reconstitute democracy. Nancy Fraser, in particular, has ar-gued strongly that in stratified societies such as ours, ". . . arrangements that accommodate contestation among a plurality of competing publics better promote the ideal of participatory parity than does a single, com-prehensive, overarching public sphere." Such "subaltern counterpublics" (blacks, women, and gays are obvious examples in the contemporary United States) create spaces in which subordinated social groups can in-vent new discourses that allow them to formulate more autonomous in-terpretations of their own identities and needs, and thus help to expand discursive space for society as a whole.[20]

Should we, in other words, aspire to build a single public realm that would seek to absorb all discursive space or is it more conducive to wider participation and democratic practice for there to exist a multiplicity of public spheres? This question will be taken up again below, but for the moment it is clear that the project of bringing back into existence an older structure—the public sphere of the 1940s or of the eighteenth century—is impossible to envision from either a moral or a practical standpoint, for it would be both indefensible in theory and futile in practice to attempt to re-form the exclusionary boundaries of the past.

Another way in which the public arena has decisively changed (perhaps one more feature of what has been thought of as its fragmentation) is the impossibility of its being reconstituted in anything like its classic form. The public sphere was born and grew with the spread of printing and literacy, and was indeed formed *around* the medium of print. One of the salient aspects of this century, though, has been the growth and prolifera-tion of electronic media—a decisive change in the means of communica-tion and representation that we may presume to be irreversible, however much its precise forms and parameters may be subject to negotiation and adjustment. In this connection, another line of criticism of the Haber-masian narrative of decline may be that such changes make necessary a reconceptualization of the public sphere, one that involves reconfiguring social communication around its new forms of mediation. Such a reconceptualization may reveal a landscape of social intercourse that, while very different from that of the print-mediated past, is too variegated and bears a too-complex relationship to the classic public sphere to be described under a simple rubric of degeneration or decline.

For the past three decades or more the cultural role of television (TV) has been the subject of heated debate, at times alternately damned as the destroyer of minds or billed as harbinger of nonlinear communication and the global village—in either case as a decisive change in cultural mediation, and one that would apparently either destroy or transform the public sphere. More recently, the revolutionary role has been assigned to internationally networked computer communication (the internet), which creates a new dimension—"cyberspace"—characterized by extremely quick access to both information and other individuals, all arranged in a nonhierarchical, "hypertextual" manner. The vision offered by cyberspace and the internet has often been related very directly to our topic: it has been hailed as offering free access by an unlimited number of individuals to a common realm of direct communication, thus promising a virtual town square, or postmodern agora, into which anyone may enter to exchange information, to debate, or simply to make social contact.

Clearly the transformative-utopian vision has informed most discussions of the internet/cyberspace, while the dystopian degeneration lament has been the dominant mode in discussion of television. But TV too has its champions, those who see in it the possibility of a transfigured, rather than a disfigured, public sphere. Over the past couple of decades scholars from a range of disciplines and methodologies, including semiotics, literary criticism, cultural studies, and cultural criticism, have approached the medium from less moralistic and less print-text-based directions, sharing the perspective that television should be studied in light of its own particular modes of engaging its audience, which obviously differ from—indeed often directly contradict—those of print. As Fiske and Hartley point out, TV is in many ways subversive of the values fostered by literacy: "The written word (and particularly the printed word) works through and so promotes consistency, narrative development from cause to effect, universality and abstraction, clarity, and a single tone of voice. Television, on the other hand, is ephemeral, episodic, specific, concrete and dramatic in mode. Its meanings are arrived at by contrasts and juxtaposition of seemingly contradictory signs and its 'logic' is oral and visual."[21]

One of the primary characteristics emphasized from the first by this line of thought has been TV's highly interactive nature: such "cool media" as television, in McLuhan's well-known characterization, ". . . are high in participation or completion by the audience."[22] Television is highly social, often addresses the viewer directly, and employs many means to create a "fictive we" that will encompass the viewers.[23] Probably this is nowhere more true than in the talk show, particularly as it has evolved over the past decade, and this has led some to see a transformation, rather than a degradation, in the contemporary crisis of the public sphere. According to Carpignano, Anderson, Aronowitz, and DiFazio:

> It is our contention that the present crisis of the public sphere is the result of, among many other factors, a crisis of legitimacy of the news as a social institution in its role of dissemination about and interpretation of events (i.e., the social construction of public life). This phenomenon is historically related to the development of new social relationships of communication embodied in the television medium that have progressively undermined the structural dichotomy between performance and audience. More than any other television genres, the new talk shows exemplify the transformation of these relationships that radically shifts the framework within which the apparatuses of mass communication and popular culture operate. They call into question the very structure of the separation between production and consumption of cultural products; they problematize the distinction between expert and audience, professional authority and layperson. Ultimately they constitute a "contested space" in which new discursive practices are developed in contrast to the traditional modes of political and ideological representation.[24]

The audience is here a direct participant in the televised event, a major player in the show, and in this role it breaks free, the authors say, from public discourses controlled by professionals, experts, and elites, and it becomes a terrain of struggle for new discursive practices. Although this appears peripheral to the revitalization of the traditional institutions of the public sphere, as well as to politics as traditionally understood, it may be that a new form of politics is emerging ("emanating from social, personal, and environmental concerns, consolidated in the circulation of discursive practices rather than in formal organizations"), which would also mean a reconceptualization of the public sphere, one in which the much-maligned TV talk show would have a place.[25]

All of this—on the unfolding possibilities within both television and internet—is highly speculative, although grounded in actual developments. One need not buy into these speculations, though, to recognize that it has at least been demonstrated that the contemporary public sphere, whatever it is, is more complex and variable than the public sphere of the eighteenth century or even of fifty years ago, and this in several ways.[26]

First, whereas the classic public sphere was print-based, the new media of the twentieth century—first film, then radio, followed by television and now the internet—have created new discursive spaces, associated with different practices, that may constitute new dimensions of the public sphere. It cannot be assumed that these changes in the modalities of the public sphere ipso facto constitute a decline.

Second, the classic public sphere was more or less severely restricted in the aspects of race, gender, and class. It is arguably the case that these restrictions have, in the course of much social struggle, eased over the course of the nineteenth and twentieth centuries, as well as in the more immediate postwar era, both de facto and de jure. Given such easements, plausible arguments can be made that the public sphere has expanded rather than contracted, even as, in other ways, it can be held that it has shrunk. Further, it cannot be simply assumed that a multiplication of competing publics represents a postmodern fragmentation, the "parcellization of a once more homogeneous public," or (in the United States) "the disuniting of America." In both cases (media and publics), their proliferation can be seen as expansion of the public sphere along new dimensions rather than as fragmentation and decline. None of this means that the public sphere may *not*, in certain respects, have atrophied or shrunk. But these growths of media and subaltern publics are not in themselves evidences of decline, and the question is a good deal more complicated than often presented, by both Habermas and most social critics.[27]

Observations along similar lines should be made with respect to the project of reconstructing a just or viable public sphere. Any theoretical reflection or practical initiative to revitalize the contemporary public arena cannot be a project simply to reverse factors that led to decline or to reconstitute what once flourished, but must involve rethinking the shape of a public sphere as it might actually exist in the contemporary world, given great changes in media and means of communication and public social interaction. A similar lesson follows from the observation that any such revitalization must create something far more inclusive than the old public sphere: in societies marked by more or less vast inequalities and diversities (and which today are not so marked?), it is very probable that the best way to accomplish this would be through building and strengthening the discourses of a multitude of counterpublics, rather than through an attempt to rebuild or enhance a unitary "official" public sphere, as the latter strategy would more likely have the effect of strengthening the power (both discursive and otherwise) of already existing hegemonic social groups.[28]

There remains, however, a possible reply to this argument, which must be considered. A distinction was drawn in the first part of this paper between the public sphere as an historical category, on the one hand, and an ideal-moral category, on the other. Through an emphasis on its character as a moral archetype, the call for a rebuilding of the public sphere and a corresponding riposte to the above can be construed in another way.

James Carey, for example, sees such calls as the summons to resurrect an ideal rather than the illusionary attempt to recreate a mythic past: "Phrases such as 'the recovery of the public sphere' . . . do not necessarily imply that there was once, long ago, in some pristine past, an era in which the public reigned. . . . The 'recovery of public life' is not an attempt to recapture a period, historical moment, or condition but, instead, to invigorate a conception, illusion, or idea that once had the capacity to engage the imagination, motivate action, and serve an ideological purpose."[29] This would seem to be close to Habermas's own intention, as he puts forward the model of the "power-free flow of communication within a single public." The contention here is that, even if such a realm did not in fact exist in the past, it was this ideal that animated the bygone public sphere, and precisely this ideal must be resurrected as an animating *telos* of our social life.

The danger in this line of approach lies in the impossibility of the proposed goal and in the practical dynamics of the relation between such unrealizable ideals and the actualities of social life that, along with social communication, inevitably takes place within and through a nexus of power relations. The hope and intention of those who advocate an impossible ethical ideal is that such a goal will draw us along a path leading ever closer to the ideal, even though it is (per hypothesis) never to be realized. The danger is that in practice such a procedure may (and, I would argue, most often does) promote blindness to existing relations of social power and to cooptation by forces of the established order. This possibility arises through the fact that social relations tend to become "naturalized"—seen as inherent aspects of the landscape. Hence those who hold power or benefit from existing structural relationships can appear as preservers of peace and order whereas those who do not must broach the issue and make their bid for greater freedom and power openly. Such outsiders are thus easily cast as disrupters of the peace and—because prevailing channels and modes of communication will tend to favor those who possess social power—villains in the quest for "power-free communication." In contemporary politics in the United States, for example, those who raise the issue of the increasingly vast unequal distribution of wealth in American society are often cast as partisans of "class warfare," whereas those who choose to pass it over in silence may take on the role of guardians of a "power-free" model of society. Habermas himself, although he would undoubtedly condemn such crass apologetics, nonetheless shows signs of being subject to this tendency, even claiming at one point that during the first part of the nineteenth century communication among voters, parties, and parliament constituted a "power-free flow of communication within a single public."[30]

Such considerations point, I would suggest, to another aspect of the way or ways in which any substantive reconstitution of the public sphere must be configured: not only must such a recomposition be inclusive with respect both to forms of media and social groups, but this public terrain must be seen, not as an impossibly power-free site, but as a particular face of the nexus of social relations. The significance of the public sphere in modern societies has not been that here we escape from the struggle for or the exercise of social power into a neutral realm of disinterested discussion, but that the public sphere has constituted an arena in which (among other things) the ceaseless contention for social space and hegemony are carried on by discursive means. If so, there are implications for the project of rebuilding the public sphere in the contemporary world. Such a project, first, cannot construe itself as simply a battle in the "realm of ideas" (although it is that), but principally as a practical social contestation. Second, this very struggle for the drastic expansion and reconstitution of the public sphere must be inextricably connected to the larger project of realigning the axes of social power in a much more equitable manner.

Notes

1. Bernard Rosenberg and David Manning White, ed., *Mass Culture: The Popular Arts in America* (Glencoe, Il.: Free Press, 1957); Bernard Rosenberg, Israel Gerver, and F. William Howton, ed., *Mass Society in Crisis: Social Problems and Social Pathology* (New York: Macmillan, 1964); Daniel Bell, "The Theory of Mass Society: A Critique," *Commentary* 22 (July 1956): 75–83; John Kenneth Galbraith, *The Affluent Society* (Boston: Houghton Mifflin, 1958).

2. Jurgen Habermas, *The Structural Transformation of the Public Sphere: An Inquiry into a Category of Bourgeois Society*, trans. Thomas Burger with the assistance of Frederick Lawrence (Cambridge, Mass.: MIT Press, 1989). The work was originally published, in German, in 1962.

3. There is obviously a connection between contemporary questions and issues involving the public sphere and those relating to public space (parallel to problems concerning the press and other institutions), but the present paper concerns only the public sphere.

4. Habermas, *Structural Transformation*; quotations are from pages 159 and 201 (in the latter I have changed the apparent misprint "whose" to "whom"). Of course, this very abbreviated adumbration neglects many facets of Habermas' rich historical analysis.

5. Many issues rear their heads here: questions about the venerable distinction between traditional and modern societies (on which see Jeffrey C. Alexander, "Modern, Anti, Post, and Neo," *New Left Review* 210 [1995]: 63–101), about the precise delineation of "bourgeois hegemony;" and about whether the category of "public sphere" is also applicable to twentieth century socialist societies, all of which I'll bypass. The terms involved are generally clear enough, and pursuing these issues would take us far too far afield.

6. Although Habermas in his later writings (*The Structural Transformation of the Public Sphere* was his first major work) has not continued a theorization and critique centered around the concept of the public sphere, several writers have pointed out that his later work, centering around the theory of communicative action, exhibits similar concerns (albeit from a less historical standpoint). See the essays by Craig Calhoun, Peter Ewe Hohendahl, and Benjamin Lee in *Habermas and the Public Sphere*, ed. Craig Calhoun (Cambridge, Mass.: MIT Press, 1992), esp. 30–32, 99–101, 402–04. A rather similar narrative of decline, but developed from different premises and with different parameters, is told by Richard Sennett in *The Fall of Public Man* (N.Y.: Norton, 1976).

7. Peter Kurth, *American Cassandra: The Life of Dorothy Thompson* (New York: Little, Brown, 1990), 319 et passim; Clarence K. Streit, *Union Now: The Proposal for Inter-democracy Federal Union (Shorter Version)* (New York: Harper, 1940), and *Union Now With Britain* (New York: Harper, 1941); Richard M. Ketchum, *The Borrowed Years 1938–1941: America on the Way to War* (New York: Anchor, 1989), 593.

8. Abbott Gleason, *Totalitarianism: The Inner History of the Cold War* (New York: Oxford, 1995), 52, 54. In counterpoint to Thomas's socialist standpoint, Republican candidate Wendell Wilkie, declaiming from the point of view of the businessman who had always seen FDR as inclined toward personal dictatorship, declared that Roosevelt's return to office would mean "an American totalitarian government before the long third term is up" (Ibid., 52).

9. James Reston, *Prelude to Victory* (New York: Knopf, 1942), 175, 229, cf. 188–89, 220.

10. Norman D. Markowitz, *The Rise and Fall of the People's Century: Henry A. Wallace and American Liberalism, 1941–1948* (New York: Free Press, 1973), chap. 2; Henry A. Wallace, "The Price of Free World Victory," *The Century of the Common Man*; and Wendell Wilkie, *One World*, both in *Prefaces to Peace* (New York: Simon & Schuster, Doubleday, Doran & Co., Reynal & Hitchcock, and Columbia University Press, 1943), 363–415, passim, and 146.

11. Irving Howe and Lewis Coser, *The American Communist Party: A Critical History* (Boston: Beacon Press, 1957), 340; Maurice Isserman, *Which Side Were You On? The American Communist Party During the Second World War* (Middletown, Conn.: Wesleyan University Press, 1982), 184. A more radical vision of the postwar world is sketched during the same period by Paul Sweezey in *The Theory of Capitalist Development* (New York: Monthly Review Press, 1968 [orig. pub. 1942]), 359–62.

12. Ruth Nanda Anshen, ed., *Beyond Victory* (New York: Harcourt, Brace, 1943), 12 passim; Lyman Bryson, Louis Finkelstein, and Robert M. MacIver, ed., *Approaches to World Peace* (New York: Conference on Science, Philosophy and Religion in Their Relation to the Democratic Way of Life, 1944 [dist. Harper & Row]), passim.

13. George H. Nash, *The Conservative Intellectual Movement in America Since 1945* (New York: Basic Books, 1976, 1979), chap. 1; John Morton Blum, *V Was For Victory: Politics and American Culture During World War II* (New York: Harcourt Brace Jovanovich, 1976), 237 and chap. 7, passim.

14. Paul Boyer, *By the Bomb's Early Light: American Thought and Culture at the Dawn of the Atomic Age* (New York: Pantheon Books, 1985), chap 3, 4, 6, 10, 19, passim. The *Reader's Digest* quote appears on p. 35. (Boyer, though, seems unaware of the extent to which the cultural themes and polarities he outlines continued the debates of the war; cf. my review of *By the Bomb's Early Light*: "A Hole in History," *The Nation*, 1 February 1986.)

15. Richard J. Barnet, *The Rockets' Red Glare: When America Goes to War: The Presidents and the People* (New York: Simon & Schuster, 1990), 204, 223–26, passim.

16. Christopher Simpson, *Blowback: America's Recruitment of Nazis and Its Effects on the Cold War* (New York: Collier Books, 1988), 228; Tom Englehardt, *The End of Victory Culture: Cold War America and the Disillusioning of a Generation* (New York: Basic Books, 1995), 120.

17. See Neil Postman, *Amusing Ourselves to Death: Public Discourse in the Age of Show Business* (New York: Viking Penguin, 1985) for a well-known example. The story I have sketched dovetails well, too, with that told by Russell Jacoby, *The Last Intellectuals: American Culture in the Age of Academe* (New York: Basic Books, 1987; Noonday Press, 1989), whose thesis is the decline to near vanishing of "the public intellectual."

18. Habermas, *The Structural Transformation of the Public Sphere*, 28–29; see also 43–51, 56. Nor has the contemporary sphere, such as it is, become *so* much more inclusive: for a detailed illustration of the way in which these historical restrictions are still at work, see Nancy Fraser, "Sex, Lies, and the Public Sphere: Some Reflections on the Confirmation of Clarence Thomas," *Critical Inquiry* 18 (1992).

19. Nancy Fraser, "Rethinking the Public Sphere," in *Habermas and the Public Sphere*, ed. Craig Calhoun, 116; see Geoff Eley, "Nations, Publics, and Political Cultures: Placing Habermas in the Nineteenth Century," in the same volume, esp. 304–06.

20. Fraser, "Rethinking the Public Sphere," 122, 122–24.

21. John Fiske and John Hartley, *Reading Television* (London and New York: Methuen, 1978), 15.

22. H. M. McLuhan, *Understanding Media* (London: Routledge & Kegan Paul, 1964), 31.

23. Fiske and Hartley, *Reading Television*, chap. 5 and 6; Robert C. Allen, "Reader-Oriented Criticism and Television," in *Channels of Discourse: Television and Contemporary Culture*, ed. Robert C. Allen (Chapel Hill: University of North Carolina Press, 1987), esp. 91–94.

24. Paolo Carpignano, Robin Anderson, Stanley Aronowitz, and William Di-Fazio, "Chatter in the Age of Electronic Reproduction: Talk Television and the 'Public Mind'," in *The Phantom Public Sphere*, ed. Bruce Robbins (Minneapolis: University of Minnesota Press, 1993), 96.

25. Carpignano, Anderson, Aronowitz, and DiFazio, "Chatter in the Age of Electronic Reproduction," 115–17.

26. A more complete discussion of the contemporary situation would have to include a fuller analysis of the much-vaunted possibilities of the internet and

cyberspace, which are probably even more over celebrated than is television over denigrated. For a beginning (deflationary) examination, see Langdon Winner, *The Whale and the Reactor: A Search for Limits in an Age of High Technology* (Chicago: University of Chicago Press, 1986), chap. 6: "Mythinformation."

27. The quoted phrases are from Jacoby, *The Last Intellectuals*, 237, and Arthur M. Schlesinger, Jr., *The Disuniting of America: Reflections on a Multicultural Society* (New York: W. W. Norton & Co., 1992).

28. The recommendation of counterpublics and subaltern spheres should not be taken to imply the fading away of an encompassing public sphere that contains them, although it does require understanding that the latter is not homogeneous, but is rather a heterogeneous setting in which, as Eley puts it, "cultural or ideological contest or negotiation among a variety of publics takes place" (Eley, "Nations, Publics, and Political Cultures," 306). See also Fraser, "Rethinking the Public Sphere," 124–27; she envisions as ideal "a society with many different publics, including at least one public in which participants can deliberate as peers across lines of difference about policy that concerns them all" (127).

29. James W. Carey, "The Press, Public Opinion, and Public Discourse," in *Public Opinion and the Communication of Consent*, ed. Theodore L. Glasser and Charles T. Salmon (New York: Guilford Publications, 1995), 373.

30. Habermas, *The Structural Transformation of the Public Sphere*, 202.

Representation, Identity, and the Communicative Shaping of Place[1]

Jean Hillier

A basic tenet to my work is my belief that local planning decisions, particularly those that involve consideration of issues of "public space" cannot be understood separately from the socially constructed, subjective territorial identities, meanings, and values of the local people and the planners concerned. Planning cannot achieve empirical reality through the work of planners alone. It is essentially intertwined with a whole range of other participants and their networks, each bringing to the process a variety of discourse types, lifeworlds, values, images, identities and emotions. I therefore explore, with relation to actor networks, ways in which different values and mind-sets may affect planning outcomes and relate to systemic power structures. By unpacking these and bringing them together as influences on participants' communication, we may come to see influences at work in decision making processes that were previously invisible.[2]

We may come to a better understanding of such processes, their shortcomings and exclusions and begin to identify new, morally inclusive, ways of proceeding. As Healey states, "The challenge . . . is to discover what the diverse people in a place are concerned about and care about, to work out a way forward which will work for most people without excluding too many interests and values."[3]

Planning practice in Western Australia is undertaken by local authorities through development control, dealing with applications for development, and strategic, or forward, planning involving developing strategies and policies for achieving desired future forms and patterns of development. Most community involvement, or public participation, takes place at this strategic level.[4]

Too often, however, limitations of resources, personnel, local knowledge, or political will result in the participation process becoming institutionalized. Grounded in reasons of inefficiency (time delays, budget blow-outs), ignorance (the "general public" lacks the "expert" knowledge to make informed comment), and stress (vulnerability to increased questions and complaints), planners set rules for the participation process. These rules, consciously or unconsciously, encourage participation by certain groups and discourage or even prohibit participation by others.

In this paper, I tell stories from the Swan Valley in Perth, Western Australia—stories from various groups of people involved in the planning decision to urbanize a semi-rural area on the metropolitan fringe. I demonstrate how individuals and groups can reinterpret place, symbols, and practices and how they can mobilize different logics to serve their purposes. As individuals, groups and organizations struggle to transform the social relations between them, they produce new "truths" by which to explain and understand themselves, their practices, and their societies.

Meanings, Values, and Identities

As Massey writes, "there are indeed multiple meanings of places, held by different social groups, [and] the question of which identity is dominant will be the result of social negotiation and conflict."[5] There can be no single reading of place. Planners have traditionally sought to "balance" readings, but have often brought to the very act of "balancing," particular mind sets that have, perhaps unconsciously, biased the scales.

Planners may also underestimate the importance of residents' attachment to their local areas and how it comprises a vital component of their social identity. A threat to their physical environment thus becomes a threat to the self. Traditional forms of planning decision making have tended to convey a message of place as identified and controlled by outsiders (the planners). Plans and policies are loaded with material, ideological, and political content that may perpetuate injustices and do violence to those meanings and identities that have not been traditionally recognized. "Instead then of thinking of places as areas with particular boundaries around, they should be imagined as articulated moments in networks of social relations and understandings."[6]

Bowles and Gintis suggest that the struggle reflects a collision between two fundamental cultural values: the expansionist logic of capitalist production and the exclusionary logic of personal rights[7]. Such values are not the property of particular groups, however, as many people have simultaneous interests in both expansion and exclusion. (Homeowners, for instance, may see benefits in increased property values from urban consolidation policies that enable them to subdivide their lots, but also want to exclude "affordable housing" from their areas for fear of depressing those same values.)

The above may be a gross simplification. Nevertheless, we do need to search for new ways and spaces of knowing that enable the expression and entrance into the debate of multiple voices, values, and representations. Such new spaces may not be comfortable for those coming from deep-seated established systems. There will inevitably be risks involved.

Citizen participation in planning epitomizes the philosophical debate between the good versus the right; the good of the local citizens in general versus the rights of the individual. Democracies need to involve communities to solve collective problems and to legitimate democratically accountable decisions. Community (the good), however, is in tension with individualism. There is a need to secure individual rights and minorities against a possible majority tyranny, but we must also be careful to avoid the opposite danger; of naturalizing existing rights and thereby "buttressing" relations of inequality.[8] The challenge for agencies of local governance is to find ways of strengthening community ties whilst protecting individuals from community oppression.[9]

Fishkin identifies deliberation as being essential for a fully communal and democratic system of decision making.[10] The social practices of discourse and communication are vital components of deliberation. It is therefore important to consider how the discourse of planning functions ideologically to shape attention and rationalize policy decisions, how it "mediates among the choices made available to us, the values we collectively espouse, and our ability to act . . . about how we should live and invest, where and with whom."[11]

In addition to the discourse of planners we need to pay attention to the ways in which other people verbalize their places in the world, their values and identities. Meanings may become more important than facts in policy deliberation. The resulting plan is "a reworking of everyday narratives to find a potentially truer, more comprehensive one. . . . Planning commands time by taking the narratives we have in mind and refashioning them."[12] I am concerned with this process of refashioning, and with the justice of the process. All our collective decisions implicate us in structures of justice and injustice; "with choice comes moral responsibility."[13]

As Walzer points out, justice requires the defense of difference; "different goods distributed for different reasons among different groups of people."[14] Justice thus becomes a "thick" moral idea, "reflecting the actual thickness of particular cultures and societies" within a "thin" set of universal principles. Justice is intrinsically related to the socially constructed idealism of the people themselves, to the things that they value and to the "personal qualities that they cultivate and mean to respect."[15]

A Communicative Way of Making Decisions

I am interested in Habermasian communicative action as a normative basis for working toward justice.[16] Empirically, however, differences in participants' discourses (language, values, images, and mind sets), have led to participants talking past each other rather than negotiating with each other. Difference has lead to polarization and entrenched positions of opposition as the stories told below well illustrate. The idea of social capital has become academically popular based on the notion that by working together in cooperation, people can generate joint benefits or "wealth."[17] To use an old cliche, the whole becomes greater than the sum of the parts.

Social capital has much in common with Habermasian communicative action. They both recognize principles of trust and mutuality-reciprocity, recognition of common ground, acknowledgment of the input of others, and the development of outputs that reflect inputs. Unlike communicative action, however, the idea of social capital values diversity.[18]

Social capital recognizes the importance of including difference and enabling difference to belong. "If the . . . system isolates people, discourages formal and informal contact, or just fails to offer the time and space needed . . . then social capital is under threat."[19]

In this paper I examine through participants' stories an understanding of the varied social constructions of meanings, values, and identities of participants in decision making processes. I then link my empirical findings back to the theory of communicative action and agree with Chantal Mouffe that "we need to make room for the pluralism of cultures, collective forms of life and regimes, as well as for the pluralism of subjects, individual choices and conceptions of the good."[20] Calling on work by both Mouffe and Seyla Benhabib, I go on to explore the reality of the "concrete other" rather than the essentially "generalized other" of Habermas and the implications of such concrete differences for the theory of communicative action.

Communication and Actor Networks

"Language is also a place of struggle."[21] A participatory communicative approach to decision making is never easy. The greater the number of participants involved, the greater the number of discourses, values, images, representations, and so on, and the greater the potential for conflict.

We should note, however, that power is intrinsic to communicative decision making. In planning terms, public participation processes may be powerful delineations of time-space in which conflicting identities are temporally and temporarily linked in discursive or dialogic contestation. As we shall see below, however, participation strategies that include hitherto excluded groups as an extension of pluralism in an attempt to achieve the public good through extending the sphere of individual rights tend to ignore the issue that some existing rights have been constituted on the subordination of others. The power structures inherent in the traditional planning system, for example, have favored elected representatives and officers of governance, who may have brought specific mind sets and identities, both of place and local resident participants, to the discussion. Such fixed images lead to nonreceptivity and result in participants talking past each other rather than engaging in open debate.

The locality itself can therefore be conceptualized as comprising layers of different outcomes over time, as actors have pursued their perceived subject positions and interests, often in competition with others. Place is a meeting point where sets of social relations (i.e., actor networks) intersect.[22]

In an actor network[23], actors in discrete situations become bound into wider sets of relations that alter the nature of their existing worlds.[24] Commitment to such networks provides forms of identity and the basis for action. In activities such as public participation, several different actor networks (including those of such non-human "actors" as aspects of nature and such non-human intermediaries between actors and networks as texts or money) will overlap and align with each other.

Constructing a new network(s) by drawing upon actors and intermediaries already in established networks, (e.g., the local authority planning system, residents' associations, etc.), the actor-network approach allows us to begin to understand how certain actors/networks are able to impose their views over those of others.

Michel Callon's actor-network theory is based on the idea that as actors struggle with each other, they determine their existence, define their characteristics, and attempt to exert themselves upon others through various human and nonhuman intermediaries.[25] Callon terms the act of an actor exerting itself upon others as "translation."[26] This process involves the four, not necessarily sequential nor mutually separate, stages of

(1) incorporation: actors join and are woven into networks.

(2) interessement: actors exert influence over others via persuasion that their position is the best one. Competing alliance are undermined.

(3) enrollment: actors lock others into their definitions and network so that their behavior is channeled in the direction desired by the enrolling actor(s).

(4) mobilization: the actor now speaks for or represents the others who have become "redefined" and passive. The representations of interest made by the lead actors are accepted as legitimate by those ostensibly being represented.

The notion of power is central to the actor-network approach, developing, as it does, Foucauldian ideas of power/knowledge. Power is regarded as the outcome of collective action. "Those who are powerful are not those who 'hold' power but those who are able to enrol, convince and enlist others into associations on terms which allow these initial actors to 'represent' all the others."[27] In so doing, they speak for the others whom they have deprived of a voice by imposing their definitions, images, and perceptions upon them. A network is thus composed of representations of beliefs, values, images, and identities, of self, others, and place. Debate and conflict occur if new representations challenge ("betray") the legitimacy of the old.

Actors will utilize whatever resources or intermediaries are available to them in order to persuade other actors to their representation or view in the pursuit of their goals. Inevitably, some actors will be able to mobilize a greater number of resources than others. In addition, although representers claim to speak for those represented, "a representation cannot capture all there is to be represented."[28] The represented or nonpresent (e.g., nature, people of lower socioeconomic status who tend not to participate, those not yet moved into the area, the unborn, etc.) may have other, unrepresented, values and aspirations. "Translation," if left to its own devices, is seldom equitable or just.

Place is therefore "shaped" by the representations of actor networks. It is dynamic; a constructed representation by actors at a particular point in time, building upon the remains of previous rounds of representation and struggle.[29]

Although it offers a useful tool for tracing and describing power relationships in decision making processes, actor-network theory may be criticized as lacking consideration of class, race, and gender issues. Similarly, it offers little understanding of the actors' ethics in their acts of persuasion (interessement and enrollment) or of the ethics and rights of nature.[30] Nevertheless, if used together with understandings of these issues above, I believe that actor-network theory does have the potential to enhance our analysis of planning decision processes.

Multidimensional Concepts of Actors, Space, and Place

In order to understand the behavior of actors in actor networks, it is important to comprehend their images/identities of where, who, why, and how they are.

The late 1980s and 1990s have witnessed the intellectual meetings of what Agnew and Duncan term the geographical and sociological "imaginations."[31] It is now generally accepted that space and place are socially constructed out of human interrelations and interactions.[32] Several authors have rejuvenated Hagerstrand's concept of space/time geography,[33] whilst space and place also figure prominently in debates around identity.[34]

Space, Place, and the State

Much has been written about the interrelationships between capital and locality and about the role of the state.[35] Corrigan and Sayer demonstrate how the ritual statements of the state define and regulate social activity. For instance, "state agencies attempt to give unitary and unifying expression to what are in reality multifaceted and differential historical experiences of groups within society, denying their particularity."[36] The authors refer to the "politically organized subjection" of citizens by the state, invoking the Foucauldian concept of normalization.[37]

A governmental preoccupation with economic growth has been imposed on the process of urban land development, and the state has tended to move beyond its traditional activities of land-use control and planning to become a major development actor engaging in urban entrepreneurialism.[38]

Places become repackaged and re-imaged to become attractive to developers and their clients. Place is a salable commodity. As such, it is necessary to remove or decrease the influence of potential obstacles to development (such as public opposition). Sujic suggests that, increasingly, "it is the property developer, not the planner or the architect, who is primarily responsible for the current incarnation of the Western city,"[39] a situation that becomes particularly worrisome when the property developer is an arm of governance.

Batten argues that governance actively uses the symbolic power of urban development to demonstrate and underline its claim to legitimacy. Governments, by demonstrating that they can "get things done," show themselves to be resolute and committed to the development of their areas. Yet, as Batten points out, although such actions may legitimize governance in the view of the metropolitan-wide public and the development capitalists, the local public who will be directly affected by the particular project may be in complete disagreement with the discourses and strategies used.[40]

There are conflicts over images and representations of the area by networks of various groups of local citizens, levels of governance, developers, and other interest groups. One or a few representations win out over others that are less acceptable to decision makers and their target group. Clashes of representation may include local residents' representations of their area as a place to live, a home, a community, a tradition as opposed to governance's images of a space to develop, to make profits, somewhere, the future. (Such different representations will be explored in detail below.) Development is thus out to work in one way or another for political ends.[41] In the process, local citizens may be losers, but few people care to look.

Can citizen networks resist the above and impose their own imagery? It is extremely difficult, but in the uncommon cases where local communities are not tamed and choose to confront development capital, the involvement of risk-taking individuals with an awareness of lobbying tactics, access to inside information and an ability to construct an alternative future that meets local needs without eliminating profit possibilities may lead to successful local resistances.[42]

The rise of the entrepreneurial state has, however, complicated the picture.[43] In instances where the state is both a major land/property holder and developer, the state is articulating its own needs and interests. As Howitt comments, "the channels of ideological power are dominated by developmentalist thinking and values in ways which discredit and marginalize alternative constructions and interpretations of emergent geographies."[44]

Attitudes such as the above, characteristic of modernist development strategies, locate urban and regional planners outside and apart from the phenomena they seek to influence. From such a vantage point, planners obtain knowledge about the objects of their attention and then proceed to control, order, and shape them as they deem appropriate. Thomas explains that a modernist approach is essentially impersonal in nature, carrying with it an absolutizing emphasis on the alleged essential similarity of all human beings, as opposed to an emphasis on their differences.[45]

Responses to Proposals for Urban Change

We need to go beyond identification of conflicts of interest to unpack the differences that influence participants' ways of giving meaning, value, and expression to tangibles and intangibles. We need to understand how the same sets of signs are read differently by different people and how people make connections between their interpretations of things and the overall ordering process of the planning system.

It should not be surprising, then, that people react to proposals for urban change in a variety of ways, not all of which are comprehensible to others involved, yet may well be patterns of action guided by deeply entrenched beliefs, norms, and values. "Social space, therefore, is made up of a complex of individual feelings and images about and reactions towards the spatial symbolism which surrounds that individual."[46] Also, it should not be surprising that many of these different representations conflict with and counteract each other.

Other important aspects influencing outcomes include structural power differences between participants and their networks that may give some people more advantage or control over others; the history of relationships between participants, including previous experiences and prior attitudes and beliefs about each other that participants bring to the discussion; and also the social environment in which participation takes place.

Planners' traditional "external" and powerful location has important implications for the understanding and discussion of claims, values, and identities. It has tended to obviate a truly participatory approach in which participants have the opportunity to enter into relationships of reciprocal respect. Instead, it has validated particular, specific forms of evidence, stressed the importance of a separation between the knower and the known, and treated personal characteristics of the knower as irrelevant. Observations are regarded as intellectual positions rather than as social constructs. Stories are irrelevant.

Planners fail to recognize that their own professional terms and definitions are themselves social constructs. Moreover, they deny themselves, as well as other participants, the opportunity to ensure that their stories are respectfully understood. Stories provide a link between private and public realms, they provide insights into meanings, behaviors, values, images and identities. "Communities cannot exist without stories."[47]

I offer below some stories as illustrations of a range of different values, representations, and identities and their influences on actors' behaviors. The stories represent a small (and inevitably biased) sample of those told in the debate over government proposals for urbanization of a semirural area in the Northeast Corridor of Perth, Western Australia (WA).[48]

These stories offer insights into the various actors' representations of themselves and their local area, embedded as they are in the tangible, physical reality of geography. They indicate a nodal point around which different identities and representations are dialogically contested, as participants' actor networks attempt to engage in interessement and enrollment of others' representations.

From these stories we gain a snapshot of an entrepreneurial state actively using symbolic imagery of place and its statutory power to overcome resident opposition to urban development. We see clashes of values and identities between the varying representations.

Stories from the Northeast Corridor, Perth, Western Australia

Perth is located in the southwest of Western Australia on the Indian Ocean coast. Urban and regional planning began in Western Australia with white settlement of an Aboriginal-occupied country in 1829. Since that time, the Perth metropolitan region has grown from a small, isolated

colonial outpost into a modern metropolis with a population forecast of
some two million people by the year 2021. Several local authorities, in
particular those at the urban fringe, such as the Shire of Swan, are experiencing increases of up to 6 percent a year, while population densities remain notoriously low. In 1990 Perth's overall urban density was just 10.8
persons per hectare, a figure comparable with that for rural Europe. Perth
thus represents a sprawling mass whose tentacles are reaching out to de-
vour the green fringes of the built-up area. The Swan Valley represents
one of these fringes in the northeast of the metropolitan area. The valley
is a delicate environment in an area of water shortage, relying on ground
water reserves. It is low-lying, with a high water table and is prone to
flooding. White residents enjoy semirural life-styles, engage in viticulture
and small holding farming, and live close to Aboriginal communities in
an area of considerable Aboriginal and white settler heritage. The practice
of settler communities living *on* the land, exploiting it for its natural resources, and using it to grow crops and for building upon contrasts
starkly with the Aboriginal concept of living *with* the land, in sustainable
harmony, a land alive with spirituality, rich with human sharing in the
past and present. Yet what represents some one hundred and fifty thousand years of history and culture is in danger of disappearing under West-
ern-influenced "civilization" and bureaucracy.

The Western Australian state government in 1990 identified a potential
population of two hundred and twenty-five thousand people for the
Northeast Corridor by 2021, an increase of some two hundred and twenty
thousand over the existing total. Although planners, in an attempt to
reach consensus, adopted a "participatory" approach to planning in the
corridor, tensions remained between the values, images, and identities of
the different participants, as indicated below.

Aboriginal Residents' Stories

The first was detailed in a journal article by I. Cunningham:

> On a recent spring day, a group of 60 Nyungahs invited senior planning
> officers from the State Planning Commission to a "very serious meeting" in
> the Swan Valley, to see important Nyungah sites and history, and how the
> sites relate to the adjoining settlement. Elders explained the Dreaming Track
> of the Waugal, sacred Avoidance Grounds, Religious Grounds, Ancient
> Meeting and Camping Places, bush foods, the inter-relatedness of everything
> in nature, and more modern cultural centres. In indigenous terms, this area
> is one of the richest in the world. Older generations, hand-in-hand with
> grandchildren, lectured on their heritage. That's how Nyungahs teach. Valu-
> able lessons are taught and learnt for the future. . . .
> A Nyungah elder stood beside Yarkiny (Turtle) Swamp, rubbing his

grandson's hair. "See our ancestors, they're all there around the swamp. See them, telling us to take care," his feeling so intense the past, present and future were brought to us all. "We want our grandchildren's children to see this, too," he said, wiping his eyes.

"We moved on to where Nyungahs lived 40,000 years ago, a site of international importance, under threat by the widening of another highway. Nearby the headless body of Yagan lies in his grave, perhaps soon to be under a highway. Across the river from here the hero Yagan, 30 pound price on his head—dead or alive—confronted George Fletcher Moore in 1833, shortly before being shot and decapitated, to ask; 'You came to our country, you have driven us from our haunts, disturbed us in our occupations. As we walk upon our own country we are fired upon by white men; why should white men treat us so?' "

"The whole land is our mother," says an elder. Nyungah ancestors were watching, all around all of us. This is spirit-land. The heritage value extends far beyond what is visible to white eyes, or what is spoken.

A senior planner was asked how he thought 80,000 people would impact on the Swan River downstream from the development. He looked puzzled: "that's nothing to do with me, you'll have to take that up with another department." He didn't think it necessary to hide his lack of interest. Nyungah religious belief and wisdom were as alien to him as the notion that straight rows of brick houses, highway pollution and removal of material bush are not progress . . .

"Just before you go," said another elder, a sick man who had made a brave and conscientious attempt to convey the significance of his heritage: "I have a question: what do you people do now? We've shown you round, explained how the creeks that flow into the river should be protected, for the sake of everybody in WA, we're trying to keep the river from dying. We've told you all our information that belongs to us, that's our precious knowledge, but we want to share it with you if it will save the river, and what do you do with our knowledge? Do you write it up? If so, we want a copy. Do you have a meeting with the Minister and tell him what we've said? What are you going to do about it? Are you going to do anything? Or are you going to leave here and just blow away like a stinking borbiny and nothing ever gets heard of today again?"

Two invaluable days in ancient Swan Valley bushland, alive with the psychic and spiritual, rich with human sharing in past and present for the future, have indeed disappeared into the bureaucratic treatment plant.[49]

The second was described in a letter written by the Nyungah Circle of Elders:

It has taken us a long time to understand and find out about all that is planned and we still do not know if we have covered everything, such as only today did we know that a section of West Swan Road marked in blue meant that it would be widened which may disturb or completely destroy the

Grave of our Ancestor, Yagan, whose head was chopped off his body and sent to England.

The Nyungah people are of a different culture and understand the land in their own way, not through white man's maps and plans.[50]

These two stories highlight the entirely different worlds of the Aboriginal people and the planning system. We see the importance of cultural argument against technical argument, the perception as alien of paper, reports, maps, and charts, and of the planning system as a whole. Memory and tradition are keys to beginning to understand Aboriginal attachment to the land. Memory is embodied in identity. There is little objective distinction between space and time.

In the tone, as well as the content of the stories, we recognize "issues, details, relationships and even people" who have been ignored and unappreciated in the past.[51] We recognize not only claims that Aboriginal people have over the land, but the importance of their self- and place-identity and "a history of betrayal and resulting fear, suspicion, distrust—which must be acknowledged, respected and addressed if working relationships are to be built."[52]

Memory is intrinsically bound up with the construction of Aboriginal identity. In the stories above, memory and traditional knowledge are being used politically. The Nyungah selected parts of their stories to tell the planners, who seemed unable to understand their importance. In similar fashion, planners select information to tell Aboriginal groups who, in turn, experience problems of being talked past rather than with.

The network of Aboriginal actors above has been completely unable to "enroll" the planners into their actor network. Despite a meeting of the two networks, Callon's stage of "interessement" was not reached. Aboriginal people failed to persuade the planners of the importance of their representation of the area. The two sets of actors failed to recognize and understand each other's intermediaries; Aboriginal sacred sites and stories and the texts, maps, and plans of the planners.

Non-indigenous Settler Residents; Stories

In the Northeast Corridor the Aboriginal communities are close to white settler residents who enjoy semirural life-styles. Many engage in viticulture, small holding farming, horticulture, and small-scale craft-related tourist ventures. Their images of, demands from, and aspirations for the area are diverse, and so, therefore, are their stories. Healy and Hillier collected quotations from a public meeting in Swan Valley:

"Our expectations [to a rural lifestyle, a green corridor] are being blighted and destroyed."

"We're concerned about water sources and wetlands."

"We will suffer traffic impacts too."

"Will we suffer financially?"

"Blight will lead to falling values; increasing poverty, crime."

"We don't know what 'urban deferred' means."

"We're not being given enough information."

"We don't all agree. Some welcome the chance to subdivide."

"We're not nimbys. We're not saying 'no' to development. We want 'appropriate development' which can be managed within environmental constraints."

"We all need to have one voice, as well as our separate voices."

"There's a different way of going about this. There must be a way of accommodating all these demands."[53]

The Henley Brook Locality Group wrote a letter to the minister for planning, an excerpt of which is below:

> There seems to be a present agenda which the bureaucrats are forcing the community into accepting. . . . It seems as though the program of community consultation was not set up to succeed . . . It seems that somewhere along the planning process the officers came to a box entitled 'community consultation' and they have run this program just so they can tick the box.[54]

These stories indicate the range of representations that residents make of the Swan Valley and Northeast Corridor as an area of history, of fragile environment, and offering a rural life-style, all of which need to be preserved as accessible to them. The identities of many are as property owners with a concern for the value of their investments. Property values and the opportunity to subdivide constitute many residents' hopes for their superannuation. Local residents view their properties as much more than units of shelter. They are financial investments, life-style symbols, social settings, and bases for leisure activities.[55] Residents recognize the diversity of their identities and aspirations, however, and that each of their different voices is valid. They are not all antidevelopment per se, but want "appropriate development" for the area, which would mean that the differing local aspirations could be satisfied.

Despite their different identities and representations the residents were united in terms of their view of the planning process as exclusive and incomprehensible and of planners as unreachable, with rigid mind sets and an unwillingness to change. Residents see themselves as powerless despite the participation process. This has exacerbated feelings of alienation and distrust of both planners and the planning system. The social space of the local residents is seen to be complex, heterogeneous, discontinuous, "and almost certainly different from the physical space in which the engineer and the planner typically work."[56]

Within the overall network of nonindigenous residents in the Swan Valley there nested several smaller, often overlapping actor networks.[57] Even within the smaller networks there were often different representations of the valley and a lack of complete agreement as to the desired outcomes. Some networks were more successful than others in enrolling planners and persuading them of their points of view.[58] The main reasons lie in their successful use of human intermediaries and the media with which they had connections.

In contrast with the Aboriginal stories above, where the planners lacked comprehension of Aboriginal texts, nonindigenous Swan Valley residents deliberately selected methods of representation (documents, public meetings, etc.) with which the planners would be familiar.[59]

Planners' Stories

Like local residents, planners are a nonhomogeneous group of people. They work at both shire and state levels. Given development proposals affecting their own residential areas, many would probably become engaged in participation processes, yet in the Northeast Corridor, they find themselves sitting on the other side of the counter, as professionals with a task to complete. My selection of stories does not encompass the views of all planners involved in the Northeast Corridor strategy and is therefore somewhat biased, and admittedly has been chosen to illustrate the key arguments of this paper. Appropriately, the planners speak through a plan:

> The structure plan is an "ideas plan" as opposed to the Metropolitan Region Scheme which is a statutory plan. The structure plan is not a legal document and has no legal force, such as the Metropolitan Region Scheme or local authority town planning schemes. The structure plan is a policy guide for the new zones and reserves currently being created by way of the major amendment to the Metropolitan Region Scheme and which, in future, will be translated into the Shire of Swan Town Planning Scheme.[60]
>
> It is in the nature of the planning process that by the time the Department of Planning and Urban Development puts out any strategic or structure plans for public comment, a great deal of homework has been done. Hopefully, in doing the homework, the Department has got its ideas right. Remembering that the Department has to take a wide range of competing interests into account in formulating its ideas, if the Department has got its ideas right by the time it publishes them for public comment there may be little room for change. If it so happens that the Department and the Government (and the local authority) believe that the Department has got its ideas

almost right and the community is of the completely opposite view then it is very unlikely that the consultation process will yield a satisfactory outcome.[61]

"Many within the community do not accept the basic proposition of a Northeast urban corridor for metropolitan expansion. Consequently, there has been very little common ground to debate any of the finer planning points or to make the most of any flexibilities which do exist in the plan. There has been a gulf between the community and the Department's respective points of view.[62]

These three stories, published in a document for a lay public readership, give a clear indication not only of tortuous plannerspeak, but also of agency identity. Planners are identifying with the agency of governance, subsumed by their institutional and professional culture.

A planner's representation of the Northeast Corridor tends to be Euclidean and instrumental. It regards the area in two-dimensional form on a map, geometrically divisible into discrete lots for the provision of housing and urban infrastructure, and as having no value in itself, "but rather its only value lies in its being 'put to work' as an instrument in the restless process of production: the 'being of things' is eclipsed by the 'doing of things'."[63] This is a view of "physical space" that, Bauman suggests, is arrived at through the "phenomenological reduction of daily experience to pure quantity, during which distance is 'depopulated' and 'extemporalized'—that is, systematically cleansed of all contingent and transitory traits," which may include Aboriginal history and sacred sites.[64]

Local citizens are regarded by the planners in the stories above as primarily motivated by a desire to preserve the status quo for themselves. They are seen as self-interested, focusing on short-term gains and losses rather than the well-being of the future population of Perth as a whole. Citizens are uninformed, often wrong and act as a hindrance to an efficient planning process.[65]

The difference between the representations of the planners and local residents may be summarized as the difference between space and place. Even so, space is not merely a passive, abstract two-dimensional arena on which things can happen. Space and place are both constructs—they are surfaces of inscription and identity, offering different meanings to different people. The inauguration of the planning system enabling the division of land into privately held and precisely demarcated lots has given planners more power/authority than local residents and therefore their interpretations of place and identity generally take precedence. Such interpretations may threaten places and identities as understood by local residents, manipulating, re-cognizing, and re-constituting them anew.[66]

Planners are experts. As Giddens points out, the difference between experts and laypersons is often a difference in power, but essentially an imbalance in skills and/or information that makes one an "authority" over the other.[67] Laypersons' knowledge, as we have seen, embodies tradition and cultural values; it is local and de-centered. Planners' expertise, on the other hand, is disembedded, "evacuating" the traditional content of local contexts, and based on impersonal principles that can be set out without regard to context—a coded knowledge that professionals are at pains to protect.[68]

Planners traditionally believe themselves to be neutral, rational, experts, offering objective and balanced appraisals rather than making value judgments.[69] Yet planners must inevitably bring their own values into their work, making judgments as to the good versus the right; what is important, which interests should carry how much weight, what is possible to be achieved, and so on. Planners and governance reserve the ultimate power to define, redefine, organize and reorganize space into a place of their choosing.

Planners, therefore, often seek to enroll other actors into their representations. Their goal is mobilization; acceptance of the planners' representations as legitimate by local residents. Public participation programs are often utilized as the means of persuasion, involving texts as intermediaries. Unfortunately, as indicated by the stories above, in the Swan Valley case, planners felt that several actors betrayed their representation. Rather than working to negotiate an outcome, the participation process degenerated into distrust and confrontation. It offered citizens only "an alienated mode of having a say in matters of public interest."[70] Participants (consciously or unconsciously) became captives of—enrolled by—the narrative content of the planning process. Feelings of personal identity were pulled in different directions as different state-sanctioned representations of meaning were put on people and their local areas. Participants' own meanings and representations were questioned and often dismissed by a proceduralist system that needs to eliminate arbitrariness and unburden the planning process from demands that are deemed extraneous to the task at hand. The result is normalized meaninglessness.

Linking Back to Theory

Participation programs that allow people to speak out, but colonize their lifeworlds in this manner, give people only "accommodative voice," whereby those in power accommodate the public by letting people have their say, although power relationships remain unchanged.[71] Simply giving other participants more voice will not work if planners do not listen to and understand their meanings and claims. There needs to be a change in the social discourse of public participation. Planners need to challenge traditional assumptions, practices, and forms of knowing and develop new practices of dialogical encounter[72] that enable communicative interaction within and between actor networks in which everyone can "justify, convince, defend, criticise, explain, argue, express (their) inner feelings and desires while interpreting those of others."[73] Understanding the representations of others and the recognition of common ground is important.

As Habermas suggests, we need to go beyond traditional knowledges and values, facts and norms, to develop a praxis of social decision making able to incorporate difference and oppositional representations and ways of knowing, and which is informed by principles of equality and justice.[74] A praxis in which people take responsibility for others, "being-for" and "being with" them.[75]

In this way, Healey suggests, planners can engage in strategic consensus-building.[76] However, is an objective of reaching consensus at the same time as respecting and valuing difference unattainably utopian? Participants in public participation programs often hold diametrically opposed views as indicated in the stories above. Can we do justice to all values, images, and identities and still negotiate consensus?[77]

I suggest not, in the traditional meaning of the term consensus, nor in the accepted Habermasian sense of the achievement of "agreements which terminate in the mutuality of intersubjective understanding."[78] We need to include in our notion of consensus, the possibility that differently formulated identities and representations may find any common links extremely precarious and that there may well be substantive and even intractable disagreements over basic issues. Mouffe argues that decisions taken in a conflictual field generally imply the repression of some representations.[79] Therefore, no consensus can exist without a "constitutive outside"—the exterior to the consensual community. I, then, prefer to adopt Love's version of con-sensus,[80] spelled with an hyphen, carrying a meaning of "feeling or sensing together," implying not necessarily agreement, "but a 'crossing' of the barrier between ego and ego, bridging private and shared experience."[81] Con-sensus involves a respect for the views of others and an attempt to understand them; "a harmony simultaneously disrupted and ordered. The contrapuntal themes are not at war or in conflict, but they come together without becoming the same."[82]

Con-sensus enables local participants in participatory decision making programs to resist the colonization of their lifeworlds by systemic power structures and to replace them with a new consciousness, an "epistemology of multiplicity."[83]). We need to reconstitute the Habermasian concept of the public sphere as one that allows for the creation of local autonomous public spheres as appropriate, to which all the public has access as relevant and which guarantees respect of differences, freedom of expression, and the right to criticism.[84]

Such public spheres would recognize a multiplicity of different meanings, ways of knowing, and forms of expression that traditionally have had no or little legitimacy in the planning process. Participants should be able to feel free to make claims on the basis of values and emotive concerns, which should be allowed equal status with technical "facts."

How does such recognition of and respect for difference between participants fit in with a Habermasian notion of communicative action? Habermas utilizes a standpoint of the generalized other, a universalist identity of the moral self such that "each individual is a moral person endowed with the same rights as ourselves."[85] By abstracting from the concrete individual identity of the other participants, everyone is able to treat all others as equal rational beings entitled to the same rights and duties as they would wish for themselves. "The moral dignity of individuals derives not from what differentiates them from all others, but from what, as speaking and acting agents, they have in common with all others."[86] Participants are accorded a definitional identity that annuls differences (of identities, values, feelings, motives, etc.) between people. Difference is something that must be transcended because it is partial and divisive, and inhibits intersubjectivity or reversibility of perspectives. Although Habermas identifies a plurality of participants' lifeworlds and expounds a need for value pluralism, he generalizes this all away in his attempt to seek some universal basis of participant equality on which communicative action can stand.[87] Participants thus become equal by bracketing difference.

Habermasian communicative action, therefore, treats different people by the same standard. Treating people equally, however, is inherently unequal (e.g., sole use of the English language in participation programs when many participants have a non-English speaking background). According to Habermas, this may be the best we can do; it represents "an attempt to exclude violence, if only to reproduce some sort of violence internally again but in a criticisable fashion."[88] But I do not think this is good enough. We can and we must do better.

I concur with the theses put forward by Seyla Benhabib and Chantal Mouffe concerning the notion of the concrete rather than the generalized other. The concrete other does not seek to abstract away from or even to tolerate difference, but instead celebrates difference as being the essence of identity.[89] It is a "pluralism that valorizes diversity and dissensus, recognizing in them the very condition of possibility, of a striving democratic life,"[90] a sentiment echoing Young's call for democratic publics to give voice to the differences within them.[91]

Recognition of concrete otherness means abandoning the reductionism and essentialism of the generalized other and acknowledging the contingent social construction of identity. It also entails accepting that the participatory decision-making process is itself part of the dynamic— identities may be reconstructed as representations are made and debated. The process "is not simply the projection of group 'interests' onto the screen of state policy, but indeed precedes this in the intricate processes of articulation through which such identities, representations, and rights claims are themselves contingently constructed."[92]

Mouffe, however, has suggested that acceptance of the idea of a concrete other threatens the very possibility of Habermasian communicative action where identity must be grounded in an unencumbered self,[93] and where antagonism, division, and conflict related to difference disappear in an all-inclusive consensus.[94] Acceptance of the concrete other accepts that some people's existing rights and identities are constituted on the exclusion or subordination of the rights of others. It therefore perpetuates a situation of marginalization and oppression.

Benhabib disagrees that this need be. She states that the standpoint of the concrete other enables participants "to view each and every rational being as an individual with a concrete history, identity and affective-emotional constitution. Our relation to the other is governed by the norm of complementary reciprocity: each is entitled to expect and to assume from the other forms of behaviour through which the other feels recognised and confirmed as a concrete, individual being with specific needs, talents and capacities."[95] She later clarifies this idea further by adding that "in assuming this standpoint, we abstract from what constitutes our commonality, and focus on individuality. We seek to comprehend the needs of the other, his or her motivations, what she searches for, and what s/he desires."[96]

It is evident that Benhabib does not regard concrete otherness as compromising universalizability of needs recognition and interpretation. In what she terms "interactive universalism" participants are able to take up the perspective of the other and develop an "enlarged mentality," a sensitivity to, and appreciation of, the wide range of moral considerations that are relevant in particular circumstances.[97]

Through communication, through dialogue with others, people can understand their differences and reach "a contested but negotiable practical understanding" of different non-assimilable ways of being and ways of knowing.[98] If communicative action, then, need not level out differences between individuals, we need to develop a form of its operationalization grounded in a transparently just procedure.[99]

Conclusions

Through telling stories from Western Australia, I have attempted to demonstrate a multidimensional concept of actors in public participation processes and an unpacking of their different representations of identity of self and place. I have tried to draw out the inherent ambiguities of the meanings and values of such representations, and I suggest the importance of a negotiated outcome to planning decision processes at a local geographical level.

In so doing I have indicated that a Habermasian based philosophy of communicative action is of value, but that it is inadequate to deal with the reality of the power of participants/actors and their representations. The shaping of public space as the result of local planning decisions is the outcome of power struggles between actor networks. I have also shown how Habermasian communicative action offers an unstable basis for consideration of difference (concrete otherness), and I have suggested that a true consensus between different participants with different, often conflicting, representations, is extremely unlikely. In its stead I offer the notion of con-sensus with its implications for participants being able to live with a planning outcome whilst not necessarily totally agreeing with it.

In answer to the question of what are the context-specific arrangements by which a plurality of interacting networks of participants, with a range of conflicting identities and representations, might be able to negotiate some form of con-sensual agreement on a local planning issue, I propose the need for a transparently just process of negotiation. My proposal is a demanding one, both geographically and philosophically, but I believe that it demands more, not less, effort to be just.

Notes

1. My thanks for helpful comments on an earlier draft of this essay to two anonymous referees and for philosophical discussion to Chantal Mouffe.

2. I acknowledge the complexity of identity and the notion of the many-sided self as espoused by M. Walzer, *Thick and Thin* (Notre Dame, Ind.: University of Notre Dame Press, 1994). In agreement with Walzer that any given society can only respond to a limited number of characteristically divided selves, I take in this paper a simplified notion of identity as represented in the planning policy process. For discussion of the construction of identities within the plurality of the social, see J. Hillier, "Cultural Values, Social Justice?" (Paper presented at Shaping Places conference, Newcastle-upon-Tyne, England, October 1996).

3. P. Healey, "Shaping Places through Collaborative Planning" (MS, 1995), 22.

4. For critical analysis of public participation strategies in Perth, see P. Healey

and J. Hillier, *Community Mobilisation in Swan Valley: Claims, Discourses and Rituals in Local Planning* (Newcastle-upon-Tyne, England: University of Newcastle-upon-Tyne, Dept. of Town and Country Planning, (Working Paper 49), 1995); P. Healey and J. Hillier, "Communicative Micropolitics: A Story of Claims and Discourses," *International Planning Studies* 1(2)(1996): 165–84; J. Hillier, "Discursive Democracy in Action," in *Planning and Social-Economic Development*, ed. R. Domanski and T. Marszal (Lodz, Poland: Lodz University Press, 1995), 75–100; J. Hillier and T. van Looij, "Who Speaks for the Poor?" in *Design for People*, ed. M. Groves and S. Wong (Sydney: PAPER, 1996), 55–66.

5. D. Massey, "The Practical Place of Locality Studies," *Environment and Planning A* 23 (1991): 267–82, 278.

6. Massey, "The Practical Place of Locality Studies," 278.

7. S. Bowles and H. Gintis, *Democracy and Capitalism: Property, Community and the Contradictions of Modern Social Thought* (New York: Basic Books, 1986).

8. C. Mouffe, *The Return of the Political* (London: Verso, 1993), 151.

9. J. Mansbridge, "Feminism and democratic community," in *Feminism and Community*, ed. P. Weiss and M. Friedman (Philadelphia: Temple University Press, 1995): 341–65, 341.

10. J. Fishkin, *Democracy and Deliberation* (New Haven, Conn.: Yale University Press, 1991).

11. R. Beauregard, *Voices of Decline* (Oxford and Cambridge, Mass.: Basil Blackwell, 1993), 5–6.

12. M. Krieger, *Advice and Planning* (Philadelphia: Temple University Press, 1981), 141.

13. Beauregard, *Voices of Decline*, 25.

14. Walzer, *Thick and Thin*, 33.

15. Ibid., 39.

16. Habermasian communicative action entails participants negotiating a consensual outcome to a problem through the force of the better argument. Ideally, all participants should speak sincerely, truthfully, comprehensibly, and legitimately and attempt to mutually understand others' points of view. See J. Habermas, *The Theory of Communicative Action*, vol. 2 (Boston: Beacon Press, 1987).

17. E. Ostrom, *Governing the Commons* (Cambridge: Cambridge University Press, 1990); R. Putnam, *Making Democracy Work*, (Princeton, N. J.: Princeton University Press, 1993).

18. Whilst Habermas accepts the plurality of participants' lifeworlds and of their voices and discourses, he retains the notion of a universal subject and appears unable to translate such plurality into his consideration of the other, continuing to rely on the concept of a generalized other as integral to his theory.

19. E. Cox, *A Truly Civil Society* (Sydney: ABC Books, 1995), 16.

20. Mouffe, *The Return of the Political*, 151.

21. b. hooks, *Yearnings* (Boston: South End Press, 1990), 146.

22. T. Marsden, J. Murdoch, P. Lowe, P. Munton, and A. Flynn, *Constructing the Countryside* (London: UCL Press, 1993).

23. Actor-network theory explores how participants coming together to discuss an issue form a network, but also how individual participants may be

members of other, wider networks of people that may or may not impinge on the debate in question.

24. J. Murdoch and T. Marsden, "The Spatialisation of Politics: Local and National Actor-Spaces in Environmental Conflict," *Transactions of the Institute of British Geographers (NS)* 20 (1995): 368–80, 378n3.

25. M. Callon, "Some Elements of a Sociology of Translation," in *Power, Action, Belief: A New Sociology of Knowledge?*, ed. J. Law (London: RKP, 1986): 196–233; M. Callon, "Techno-economic Networks and Irreversibility," in *A Sociology of Monsters,* ed. J. Law (London: Routledge, 1991).

26. Callon, "Techno-economic Networks and Irreversibility."

27. Murdoch and Marsden, "The Spatialisation of Politics," 372.

28. Marsden, Murdoch, Lowe, Munton, and Flynn, *Constructing the Countryside*, 31.

29. It is impossible to conceptualize a diagram of actor networks over time. I offer the metaphor of an infinitely expanding Rubic's cube, wherein each small segment represents an actor, interacting with the others it touches. Through time, the cube may be turned partially or entirely. The faces of some segments will remain touching in their previous "networks," whilst others will form new networks. New patterns and relationships of networks are formed with every twist of the cube, while the sediments of some old patterns (or their influence) may still be retained.

30. These particular issues of ethics of persuasion and the ethics and rights of nature are discussed in detail in J. Hillier, "Going Round the Back? Complex Networks and Informal Associational Action in Local Planning Processes." (paper presented at the ACSP conference, Florida, November 1997); and J. Hillier and T. van Looij, "Pragmatic Planning and the Representation of Nature" (paper presented at the Environmental Issues Conference, Melbourne, October 1997).

31. J. Agnew and J. Duncan, "Introduction" in *The Power of Place*, ed. J. Agnew and J. Duncan (London: Unwin Hyman, 1989), 1–8.; J. Duncan and D. Ley, ed., *Place, Culture, Representation* (London: Routledge, 1993); M. Keith and S. Pile, ed. *Place and the Politics of Identity* (London: Routledge, 1993).

32. Place involves the appropriation and transformation of space, which is a more neutral concept. Place-making comprises the physical creation and de-creation of landscape and the socioeconomic mediation of a "felt sense of the quality of life." A. Pred, "Structuration and Place: On the Becoming of Sense of Place and Structure of Feeling," *Journal for the Theory of Spatial Behaviour* 13 (1983): 45–68, 58.

33. For example, D. Massey, "Politics and Space/Time," in *Place and the Politics of Identity*, ed. M. Keith and S. Pile, 141–61.

34. See A. Pred, "Place as Historically Contingent Process: Structuration and the Time Geography of Becoming Places," *Annals of the Association of American Geographers* 74 (1984): 279–97, 279.

35. See, for example, work by D. Harvey, *The Urbanisation of Capital*, (Oxford and Cambridge, Mass.: Basil Blackwell, 1985); D. Harvey, *The Condition of Postmodernity* (Oxford and Cambridge, Mass.: Basil Blackwell, 1989); D. Harvey, *The Urban Experience*, (Oxford and Cambridge, Mass.: Basil Blackwell, 1989);

D. Harvey, "From Managerialism to Entrepreneurialism: The Transformation of Urban Governance in Late Capitalism," *Geografiska Annaler* 71B no.1 (1989): 3–17; D. Harvey, "From Space to Place and Back Again: Reflections on the Condition of Postmodernity," in *Mapping the Futures*, ed. J. Bird, B. Curtis, T. Putnam, G. Robertson and L. Tickner (London: Routledge, 1993): 3–29; D. Harvey, "Class Relations, Social Justice and the Politics of Difference," in *Principled Positions*, ed. J. Squires (London: Lawrence and Wishart, 1993): 85–120.; D. Massey and J. Allen, ed., *Uneven Re-development*, (Milton Keynes: Open University Press, 1988); D. Massey, *Spatial Divisions of Labour*, (London: MacMillan, 1994); Neil Smith, *Uneven Development*, (Oxford and Cambridge, Mass.: Basil Blackwell, 1984); N. Thrift, "Localities in an International Economy" (paper presented to ESRC Workshop, UWIST, Cardiff, Wales, 1986); N. Thrift, "The Geography of Late Twentieth Century Class Formation," in *Class and Space*, ed. N. Thrift and P. Williams, (London: RKP, 1987): 207–53; N. Thrift, "Introduction: New Models of the City," in *New Models in Geography*, vol. 2, ed. R. Peet and N. Thrift, (London: Unwin Hyman, 1989): 43–54.

36. P. Corrigan and D. Sayer, *The Great Arch* (Oxford and Cambridge, Mass.: Basil Blackwell, 1985), 4.

37. Ibid., 7.

38. Harvey, "From Managerialism to Entrepreneurialism," 3–17.

39. D. Sujic, *The 100 Mile City* (London: Andre Deutsch, 1992): 34–35.

40. D. Batten, "The Port Melbourne Condition: Normative Geographies of Legislation for Urban Development" (Ph.D. Disc., University of Melbourne, 1994), 174.

41. Sujic, *The 100 Mile City*, 84.

42. See, for example, empirical research by J. Hillier and G. Searle, 1995. *Rien ne va plus: Fast Track Development and Public Participation in Pyrmont-Ultimo, Sydney* (Sydney: UTS, UTS Papers in Planning, no. 3, 1995).

43. Harvey, "From Managerialism to Entrepreneurialism."

44. R. Howitt, "SIA, Sustainability and the Narratives of Resource Regions: Aboriginal Interventions in Impact Stories" (Macquarie University, School of Earth Sciences, Sydney, 1994), 4.

45. H. Thomas, "Introduction," in *Values and Planning*, ed. H. Thomas (Aldershot: Avebury, 1994):1–11, 34.

46. D. Harvey, *Social Justice and the City*, (London: Edward Arnold, 1973), 34.

47. D. Maines and J. Bridger, "Narratives, Community and Land Use Decisions," *The Social Science Journal* 29, no.4 (1992): 363–80, 366.

48. For more detail of the debate, see Hillier, "Discursive Democracy in Action"; Healey and Hillier, *Community Mobilisation in Swan Valley;* Healey and Hillier, "Communicative Micropolitics." Research into many more such stories in the area is ongoing (Australian Research Council grant A79602576).

49. I. Cunningham, "Our Irreplaceable Heritage," *The Western Review* (November 1994): 7–8.

50. Excerpts from a letter from the Nyungah Circle of Elders to the WA Minister for Planning, 24 April 1993, 3.

51. J. Forester, *Beyond Dialogue to Transformative Learning: How Delibera-*

tive Rituals Encourage Political Judgement in Community Planning Processes (Haifa, Israel: Centre for Urban and Regional Studies, working paper, 1993), 31.

52. Ibid., 31.

53. Quotations from public meeting, reproduced in Healey and Hillier, *Community Mobilisation in Swan Valley.*

54. Hillier, "Discursive Democracy in Action," quoting Henley Brook Locality Group letter to Minister for Planning, 26 February 1992.

55. Healey, *Shaping Places through Collaborative Planning.*

56. Harvey, *Social Justice and the City*, 35.

57. See Hillier, "Discursive Democracy in Action," for identification of locality groups, interest groups, etc.

58. The "Friends of the Valley," a group located on the east bank of the River Swan, has managed successfully to fend off virtually all proposals for urbanization in its area of representation.

59. See Healey and Hillier, *Community Mobilisation in Swan Valley*; Healey and Hillier, "Communicative Micropolitics."

60. Department of Planning and Urban Development, *North East Corridor Structure Plan, Final,* (Perth: DPUD, 1994), 4–5.

61. Ibid., 20.

62. Ibid., 20–21.

63. P. Hoggett, *Partisans in an Uncertain World: The Psychoanalysis of Engagement* (London: Free Association Books, 1992), 107.

64. Z. Bauman, *Postmodern Ethics,* (Oxford and Cambridge, Mass.: Basil Blackwell, 1993), 145.

65. See also, T. Tyler, K. Rasinski, and E. Griffin, "Alternative Images of the Citizen: Implications for Public Policy," *American Psychologist* 41, no. 9 (1986): 970–78.

66. Whilst the above may often be true of some communities, there are, nevertheless, more outspoken communities whose members use manipulative stories to direct outcomes in favour of their values and identities. Noise and hassle may be sufficient may be sufficient for decision makers to "lose their nerve" and vote to appease their constituents, while long delays may cause applicants to withdraw proposals.

67. A. Giddens, "Living in a Post-Traditional Society," in *Reflexive Modernisation,* ed. U. Beck, A. Giddens and S. Lash, (Cambridge, Eng.: Polity Press, 1994): 56–109, 84.

68. Ibid., 85.

69. See C. Hoch, *What Planners Do,* (Chicago: APA, 1994).

70. Habermas, *The Theory of Communicative Action*, 370.

71. E. Sampson, "Identity Politics: Challenges to Psychology's Understanding," *American Psychologist* 48 (1993): 1219–30.

72. R. Bernstein, *The New Constellation,* (Cambridge, Eng.: Polity Press, 1991), 337.

73. S. Chambers, "Discourse and Democratic Practices," in *The Cambridge Companion to Habermas,* ed. S. White (Cambridge: Cambridge University Press, 1995): 233–62, 242.

74. Ibid.

75. Bauman, *Postmodern Ethics.*

76. P. Healey, *Strategic Spatial Planning as a Process of Argumentation,* (Perth: Curtin University, Centre for Architecture and Planning Research, occasional paper 23, 1994), 5.

77. For issues of entrenched positions, lobbying, etc., see Hillier, "Going Round the Back?"

78. J. Habermas, *Communication and the Evolution of Society,* trans. T. McCarthy (Cambridge: MIT Press, 1979), 3.

79. C. Mouffe, *Dimensions of Radical Democracy, Pluralism, Citizenship and Community,* (London: Verso, 1992), 235.

80. N. Love, "What's Left of Marx?" in *The Cambridge Companion to Habermas,* ed. S. White (Cambridge: Cambridge University Press, 1995): 46–66, 62.

81. N. Holland, *Five Readers Reading,* quoted in *Womens' Ways of Knowing: The Development of Self, Voice and Mind,* M. Belenky et al., (New York: Basic Books, 1986), 223.

82. Love, "What's Left of Marx?", 62.

83. G. Gomez-Pena, 1993. *Warrior for Gringostroika* (St. Paul, Minn.: Greyworld Press, 1993), 38.

84. On forums, arenas, and courts for possibilities of operationalizing such ideas, see J. Bryson and B. Crosby, *Leadership for the Common Good* (San Francisco: Jossey-Bass, 1992).

85. S. Benhabib, *Situating the Self,* (Cambridge, Eng.: Polity Press, 1992), 10.

86. Ibid., 151.

87. J. Habermas, *Justification and Application,* trans. K. Baynes (Cambridge: MIT Press, 1993); J. Habermas, *The Past as Future,* (interviews with M. Haller), (Cambridge: Polity Press, 1994).

88. J. Habermas, *Postmetaphysical Thinking,* trans. M. Hohengarten (Cambridge, Eng.: Polity Press, 1992), 479.

89. S. Benhabib, *Critque, Norm and Utopia* (New York: Columbia University Press, 1986); S. Benhabib, "Communicative Ethics and Current Controversies in Practical Philosophy," in *The Communicative Ethics Controversy,* ed. S. Benhabib and F. Dallmayr (Cambridge: MIT Press, 1990): 330–69; S. Benhabib, "In the Shadow of Aristotle and Hegel: Communicative Ethics and Current Controversies in Practical Philosophy," in *Hermeneutics and Critical Theory in Ethics and Politics,* ed. M. Kelly (Cambridge: MIT Press, 1990): 1–31; Benhabib, *Situating the Self;* C. Mouffe, "Radical Democracy: Modern or Postmodern?" in *Universal Abandon?,* ed. A. Ross (Minneapolis: University of Minnesota Press, 1988), 31–45; Mouffe, *Dimensions of Radical Democracy;* Mouffe, *The Return of the Political;* C. Mouffe, "Post Marxism: Democracy and Identity" in *Environment and Planning D, Society and Space* 13 (1995): 259–65.

90. Mouffe, "Post Marxism: Democracy and Identity," 265.

91. I. M. Young, *Justice and the Politics of Difference,* (Princeton, N. J.: Princeton University Press, 1990).

92. Mouffe, *Dimensions of Radical Democracy,* 121. Links with actor-network theory are clear.

93. Mouffe, "Radical Democracy: Modern or Postmodern?"

94. Mouffe, *Dimensions of Radical Democracy*; Mouffe, *The Return of the Political*, 140–42.

95. Benhabib, *Critque, Norm and Utopia*; Benhabib, "Communicative Ethics and Current Controversies," 341.

96. Benhabib, *Situating the Self*, 159.

97. Ibid., 165. Unacceptable morals would be dismissed in communicative action by the force of the better argument of acceptable morals.

98. J. Shotter, *Cultural Politics of Everyday Life* (Buckingham, Eng.: Open University Press, 1993), 116.

99. I have developed a theoretical account of a procedurally just approach to communicative decision making that incorporates concepts of difference in J. Hillier, "Beyond Confused Noise?" (paper presented at In Search of the Public conference, Fremantle, December 1996).

Maps and Entitlement to Territory

Mario Pascalev

In this paper, I examine the possibility that geographic maps supply evidence of traditional occupation of an area by a nation. Traditional occupation is often an important consideration in legal, moral, or political claims to territory. Ordinarily, maps are considered "scientific abstractions of geographic reality." As such they supply the principal evidence in support of claims of traditional occupation that aim at securing historical title to territory. I argue that, at least with respect to ethnic cartography, the claim that geographic maps are objective and scientific tools of representation fails. Rather than being objective and impartial, maps project nationally colored visions of the territorial extent of nations. The fitness of maps as vehicles for the national imagination lies in their concrete visual presentation. They are capable of winning both national and international support for a certain territorial vision. However, although maps can create political support, they cannot uphold a moral claim to the territory under consideration.

Maps are often used to acquire political support for a territorial design although they are purported to back a moral claim to said territory. Such a strategy appears intrinsic to the process of nation-building. I substantiate my observations by examining the mapping of Macedonia in the nineteenth and twentieth centuries. Although the chronicles of ethnographic mapping of Macedonia reflect the nation-building efforts of Albanians, Bulgarians, and Greeks, among others, one endeavor deserves particular attention: the maps of the prominent Serbian geographer Jovan Cvijic. Cvijic's endeavor is remarkable for its success in advancing national claims and completely reversing the established international perceptions in a period of only twelve years; thus, I will illustrate my claims with his work.

There are a number of principles governing claims to territory in international law and international morality. Some of these are conquest (effective control and administration of a territory by a state), acquiescence (abandonment of a territorial claim through inaction), cession (transfer of title via state treaty), rectification of past injustices, and traditional occupation.[1] The first, conquest, was once the most prevalent mode.[2] However, it is almost entirely rejected today. The next several principles are appropriately employed in only limited circumstances. The last, traditional occupation, appears to be a sound and fair principle of acquisition. It holds that if a national group has occupied a territory longer than any other group, the state that comprises this group has a right to the territory.[3] There are theoretical problems with this principle but discussion of them is beyond the scope of this essay. It is nonetheless important to stress that traditional occupation is widely perceived as a fair principle of acquisition of territory.

The argument that grounds a particular state's right to territory from the principle of traditional occupation proceeds as follows:

1. A state is entitled to the territory that representatives of its nation(s) traditionally occupy.
2. Traditional occupation is an objective geographic fact about nations.[4]
3. Objective geographic facts about nations are represented in objective and impartial ethnographic maps.[5]

If it is the case that

4. An objective and impartial ethnographic map represents a certain nationality A as occupying a territory B,

From the above premises, we can deduce the conclusions of the argument:

5. Nation A in fact occupies territory B (follows from premises 2, 3, and 4).
6. The nation-state of A is entitled to territory B (follows from premises 1 and 5).

A fact about traditional occupation will be represented on ethnographic or other maps. These may be historical, or may result from contemporary research. In any case, a moral claim to territory grounded in traditional occupation will be mediated by maps. The maps will, thus, impact on international law and morality. I will examine whether such influence is justified. In doing so, I will turn to analysis of the position that holds that maps are legitimate contributors to international moral claims (expressed in the above argument).

A number of premises constitute the position that maps legitimately contribute to international moral claims to territory. I shall not treat all of them at length, but I want the reader to be aware of them. On the one hand, the position includes claims related to entitlement to statehood and territory, such as the thesis that all nations and only nations should become states, as well as claims that a procedure, such as traditional occupation, creates a title to territorial jurisdiction for nation-states.[6] I will not question this notion of entitlement, as such a critique would go deep into the core of Nationalism, and I need not make the effort to refute the Nationalist position as such. On the other hand, the position includes a number of assumptions that nations are distinguishable on the basis of objective properties, and that their differences could be scientifically represented. In the rest of this section, I will concentrate on the way in which nations are objectively distinguishable and the meaning of such distinguishability for traditional occupation. I will not question the character of the notion of entitlement but, rather, just the possibility of its application.

It should be noted that the claim that geography is objective and impartial has been criticized in the past. David Harvey takes issue with this claim from the Marxian view that ideology is a "false consciousness" reflecting class bias.[7] He claims that geography too has a strong ideological content.[8] Others like J. B. Harley criticize the claim that geography is objective and impartial, building on the work of Foucault and Derrida.[9] These are fine works that make interesting points. However, their critical conclusions with regard to the objectivity and impartiality of geography are unavailable to someone who does not accept the large number of assumptions that underpin these conclusions. In the case of the Marxist geographers, the assumptions are about the existence of laws of history, class struggle, and the like. In the case of the postmodern geographers, the assumptions are about the embeddedness of power in language and the perception of the social world as a text.

In contrast, the project of this paper is a limited one, something not uncommon for analytic moral philosophy. I would like to show that geography fails to be objective and impartial for one particular reason: the attempt of human geographers to produce objective and impartial maps of the dwellings of different nations cannot help but fail. I avoid making assumptions about the metaphysic of the social world or of history. Thus, my analysis is available to researchers coming from a variety of paradigms. They need not subscribe to grand theories in order to accept my results.

I.

Two beliefs contribute to the success of maps in international affairs. First is the belief that nations are objective entities, distinguishable according to reliable and objective criteria; the existence of nations is a matter of fact (the distinguishability thesis). Second is the belief that maps are scientific representations of established facts about an undisputed objective reality (the representation thesis).[10] Scale, colors, lines, and categories bear strict, unique correspondence to these undisputed facts. I will argue that the above two theses are groundless and that the argument, thus, fails.

First, let us consider the distinguishability of nations thesis, that nations are entities, distinguishable according to objective criteria. Notice that this thesis involves four claims: (1) There is a property or a set of properties that define nationhood; (2) These properties are objective; (3) If something is a property or a set of properties of a nation, then it is a criterion for distinguishing between different nations; (4) These properties apply universally and are universal distinguishing criteria. The distinguishability thesis could be defended on different levels, depending on the thickness of assumptions concerning the nature of nations. From one extreme, a very thick notion would have it that a nation is an objective entity, that precedes and determines the outlook of its members according to its purpose.[11] On this account, nations would evidently differ as natural kinds. From the other extreme, one might argue for a much thinner notion of a nation. Whatever nations are, they have properties that uniquely and objectively separate them from other nations. The former account is so laden with metaphysical assumptions, that virtually no one today would subscribe to it. It is also more difficult to defend. A charitable criticism should deal with the latter account, which is more defensible.[12]

There is nothing striking about the refutation of the distinguishability thesis that I am going to offer. I begin by asserting that there is not a single defining property, or a distinguishing criterion of nations that is common to all nations. I, thus, claim that even if we grant the existence and objectivity of properties of nationhood, we have to deny the fourth claim above, that these properties apply universally. A review of the most discussed properties of nations, such as language, ethnicity, culture, or religion, shows that they all fail in some instances to be properties of nationhood. This is a familiar criticism of the concept of a nation. However, this criticism against the distinguishability thesis is not the last word. A familiar reply is also offered: Nations need not share one defining property or set of properties. Rather, they need share only a family resemblance of a set of properties.[13] Thus, if, for instance, religion, ethnicity, and language are among these defining characteristics, some nation's identity may be founded on common language and ethnicity, another nation's identity may be based on a common religion and language, and a third one's identity grounded in shared ethnicity and religion.

Let us accept for the sake of argument that the family resemblance theory is correct. That is, I grant that there are defining properties of nations *and* that they are criteria for distinguishing different nations in a family-resemblance fashion. Nevertheless, this may not be enough to save the distinguishability thesis. In order for a family resemblance matrix to work, it must be objective. That is, there must be a matrix that *ex ante* assigns the properties relevant to the nationhood of each particular nation; else, subjectively formed matrices may contradict each other. A partition of the earth's surface according to the languages spoken will differ from a partition based on religion or ethnicity, or on different combinations of these. Therefore, maps reflecting different criteria will contradict each other. In sum, if no universal criterion for nationhood is possible, then there must be an objective family-resemblance matrix, preordaining the defining elements of each nation. The existence of such a matrix is extremely dubious, as established nations seem to lose characteristics and incorporate new characteristics with the passage of time. If this is the case, then the claims of science concerning nations cannot fail to assume an *ad hoc* form.

Perhaps, we should imagine the objectivity of a family resemblance matrix in a less strict way. Rather than assuming that such a matrix will order properties in every particular case, we may think of it as assigning priority of certain properties, if, and when, they apply. For instance, such a lexicographic ordering may assign first priority to language, second priority to religion, and so on. If, for example, language were the attribute of nationhood for a particular nation, and not for a second nation; then language would be applied as a distinguishing criterion, even to the detriment of the second nation.

This refinement does not help the family resemblance theory a whole lot. The application of a lexicographic ordering of characteristics of nationhood should apply not only in new situations, but also in general, including for revisions of existing nations. In my example from the previous paragraph, Germany should be able to incorporate Austria, Alsace, and parts of Switzerland. Likewise, different revisions would be prompted by applying various other orderings. I submit that it is counterintuitive and, perhaps, impossible to require the dismantling or the handicapping of established nations. If there is an argument in favor of a certain particular ordering, it cannot be used to override existing arrangements, as it could only refer to the value of nations, the same value that would protect the existing nations. To make things even worse for the family resemblance theory, my objection to it from the previous paragraph also applies to this construal of it. The defining characteristics of nations change over time. If it happens that a nation adopts a low priority characteristic of nationhood instead of a former high priority one, this nation risks revision of its population and territory.

Finally, the way in which we conceive of the objectivity of family resemblance of characteristics may be completely different. Namely, it could be determined *ex post*. Suppose that all characteristics belonging to the family resemblance set are employed in multi-dimensional coordinates. It may be conjectured that when we graph all nations' characteristics on the respective coordinate, each nation will form a distinct cluster. What it would take to refute this model is an existential claim that nations do not cluster in the suggested way. For instance, Mexicans will cluster together with some Americans, particularly with the Mexican Americans. And this is not the only example. It may turn out that some Belgians cluster closer together with the French than with their fellow-Flemish.

Thus, I shall conclude that although nations may be defined in terms of a family resemblance of facts, such facts cannot be used in an objectivist geography and only such a geography would be acceptable as impartial and objective science.[14]

The second thesis—that the maps provide strict representation of facts about nations—is shaken by the rejection of the first thesis. The appeal of science in general, and geographic science in particular, comes from the belief that it interprets objective facts in a universal and impartial way. If there are no objective facts (universal, or according to an objective matrix) about nations, then strict representation of nations is impossible. In its place, I suggest that what geographers do is construct nations. Independently, even though we may grant for the sake of the argument that there are such universal facts, maps are still dependent on the categorization, generalization, and sub-division of facts.[15] These are all in the hands of political geographers and their sponsors. If the geographers who perform these functions happen to be partial, nothing would prevent them from manipulating the tools at their disposal. I suggest that even if one does not accept my thesis that geographers by necessity construct aspects of nations rather than represent objective facts about nations because no such facts are available, one may appreciate the constructive work of geographers despite the background of objective facts in a particular case.

The success of the work of national geographers on maps is due to the wide acceptance of the divisibility and representation theses. As we saw, the latter theses turned out to be false and the geographers' work was stripped of the pretense of objectivity and impartiality. In spite of, or perhaps just *because of* the failure of objectivity and impartiality, the work of geographers has an important impact on national identity. There are two aspects of this impact. First, geographical ideas, inscribed with colors and lines on reproducible images, become a part of national imagination. Fellow nationals have a spatial insight about the location, size, and limits of their nation. Maps become carriers of transgenerational memory; they link the imaginations of different generations. Maps seal the imprints with the air of tradition, longevity, and authenticity. They become the spatial memory of a nation and may even become programs for action. Second, maps influence international politics by giving political aspirations the form of a moral claim.

II.

The mapping of Macedonia by Cvijic constitutes a particularly articulate example of using maps for directing a national imagination and, at the same time, contriving a legal and moral claim on the grounds of traditional occupation. The case study will highlight problems with the assumptions of objectivity and impartiality of geography, pinpointed in the earlier sections of this paper. I will outline several consecutive stages of the ethnographic cartography of Macedonia. I will use extensively a remarkable, and yet nearly forgotten, study by Henry Wilkinson, as well as the accessible original sources.[16]

Hellenistic Maps

The earliest descriptions of European Turkey reflected the wide-spread belief that the Greeks were the major inhabitants of the Balkans.[17] These are the Greeks proper ("Hellenes") and Greek-related tribes, such as the plebeian Greeks ("Pelasgians"), and the original inhabitants of the peninsula, the "Illyrians," some of whom are preserved (Montenegrins), some are romanized (Vlachs) and still others Slavized (Bulgarians).

The most important factor that brought about this belief is the importance ascribed to religion for nationhood. Belonging to the Greek Orthodox Church was considered a sufficient condition for being a Greek. Note that all "Illyrians" (i.e., Bulgarians, Montenegrins, and Vlachs) are of Orthodox persuasion. There were certain reasons for this type of perception. The Millet system of religious freedoms in the Ottoman Empire gave religious autonomy to Moslems, Orthodox, and Jews. However, the Greek patriarchy was recognized as the sole religious leader of the Orthodox Christians. This may have produced the impression that the Orthodox population was mostly Greek. An additional factor that may have led to the perception that the Greeks are the major inhabitants of the Balkans is that Western-Europeans familiar with antiquity longed for the restoration of Byzantine Greece and Ancient Greek culture. Also, Greek sailors and merchants sailed the seas and had a high profile compared to the inland populations.

Slavonic Maps

The pro-Hellenic sentiment produced a vastly distorted ethnographic portrait of the Balkans, running against the self-image of the great bulk of the populace. This distortion affected primarily the Slavs, who constituted the overwhelming majority of the population. As a consequence, the next generation of maps of the Balkans acquired the form of a revision from the Slavic point of view. The defining fact about nations used in these maps was language. The most important maps of this period were produced by P. J. Safarik in 1842, Ami Boué in 1847, and G. Lejean in 1861.[18] These maps afforded the bulk of the peninsula to Bulgarians and Serbs. Macedonia on these views was homogeneously inhabited by Bulgarians.

The most important reason for this influential shift in perception is the heightened role of language as a property of nationhood. The Romantic movement awoke the interest in original languages and folk traditions. Notably, Safarik was one of the first researchers of the languages and folklore of the Slavs, as well as a champion of Pan-Slavism. Pan-Slavic ideas spread widely among the Slavonic populations in southern, central, and eastern Europe. This ideology rose from feelings of brotherhood and common destiny. The successful struggle of the Serbs and Bulgarians for religious and political freedom was highly visible. After the massacres of 1876 following the April Uprising, Europe's sympathy was on the side of Bulgarians.

A notable feature of the maps of 1842–70 is that the delineation between Bulgarian and Serbs is clear and undisputed according to all authors. It ran northwest of the line Nis-Skopje. Similarly, a map produced in Belgrade by Davidovic in 1848 agreed with Boué on the southern limits of the Serbian population. We could attribute this harmony of opinions to the spirit of camaraderie brought about by Pan-Slavism. Additionally, the national aspirations of the Serbs were at the time directed to the Adriatic to encompass Serbo-Croatian populations in Bosnia and Dalmatia. Serbs imagined a country spreading west and southwest, rather than one spreading south.

As a result of the cartographic activity of the Slavophiles, a moral claim of the Bulgarians to Macedonia was established. When the circumstances ripened, this claim justified political arrangements favorable for Bulgaria. The struggle for Bulgarian religious independence resulted in the creation of the Bulgarian Exarchate, the borders of which were established by Turkish plebiscites. Not incidentally, the borders of the Bulgarian Church coincided with the findings of Safarik's and Boué's maps about the distribution of the Bulgarian language. Later, the Conference of Constantinople attempted to afford autonomy to Bulgaria within the borders of the Exarchate. The conference, for the first time, revealed to the Balkan people the power of the maps. As Wilkinson puts it:

> If the Conference did nothing else, it did create interest in ethnographic maps because it became clear that through the medium of ethnographic ideas the Bulgarians had gained moral ascendancy over all other people of the Balkans . . . [it was] due to the simple fact that for thirty years the greater

part of the territory between the Danube and the Aegean, between the Macedonian lakes and the Black sea, had been colored as Bulgarian on scores of ethnographic maps . . . In the simple flat colors were to be seen the hopes and aspirations of a nation.[19]

Greek Maps

The realization of the strength of the Slavic element on the Balkans and the possible political implications, e.g., a large Bulgarian state dominating the peninsula, raised concerns on both a general and a local scale. The Greek public was late to apprehend the power of the maps. It watched the struggle for Bulgarian independence with growing irritation and almost fear.

When the Greeks finally registered the important changes in public perception with regard to the ethnic composition of the Balkans, and Macedonia in particular, they tried to make up for the lost time. The most important Greek map was published in 1877 by E. Stanford.[20] This map has two interesting features: (1) It rejected language as a criterion of nationhood, although language was by then broadly perceived as a "natural" and fundamental criterion. Instead, it introduced the criterion of cultural and historical bonds. (2) It created and applied new mixed categories, in a demonstration of the constructivist capacity of geographers. The Greek maps extended the Greek ethnographic frontier as far north as the Balkan mountain range. The Vlachs were classified as Greeks on this map, and broad territories were shown as mixed Turk-Greek, Bulgarian-Greek and Albanian-Greek. Astonishingly, many Slavs were pronounced "Bulgarophone" Greeks. Some support for the Greek maps came from the Ottoman statistics, which tended to register all Orthodox Christians as Greeks.

The dating of the pro-Greek map of Stanford clearly betrays it as a reaction to the political developments of the time. Overall, the map was perceived by the geographic community and the international public as exaggerated, partial, and lacking scientific value. In an apparent reference to Stanford's map, English ethnographer Sir A. J. Evans scornfully remarks:

> One of the most comic attempts of this competitive ethnography was a map published some years ago under Athenian auspices . . . According to this Macedonia was, for practical purposes, divided into two elements—The Greek and the "Bulgarophone Greeks"—as if some Celtic enthusiast should divide Britain between the Welsh and the "Anglophone" Welsh![21]

Austrian Maps

Austrian projects for the Balkans prompted their participation in the war of the maps. The general course of their participation, again as in the

Greek case, was to undermine the Slavic cause. However, Karl Sax's map of 1877 was much more subtle than Stanford's map.[22] Rather than implausibly dismissing language as a criterion of nationhood, Sax suggested that it is a necessary but not a sufficient condition for the representation of the diverse nature of the Balkan nations. He proposed that the relevant criteria of nationhood consist in both language *and* religion. As a result, instead of the four prominent nations (Albanians, Bulgarians, Greeks, and Serbs, as well as the minority Turks and Vlachs), corresponding to the six languages spoken on the peninsula, Sax introduced *twenty-eight different nationalities!*[23] It seems that although increased accuracy is laudable, it is subject to the requirements of scale. Putting a group of several hundred people, e.g., the Bulgarian Uniates, on a large-scale map would cause them to appear much more significant than they are proportionally. It seems that, inevitably, a geographer's decision for or against more precision would favor some of the disputants.

Nonetheless, Sax's message was straightforward: The situation on the Balkans and, particularly, in Macedonia is very complicated. Therefore, simple solutions like the available maps representing four to six nationalities are inadequate. Likewise a simple political solution, such as granting each Balkan nation statehood, could not be a feasible solution for the problems of the peninsula. It is unclear what general political suggestions Sax would have had for the kind of situation that was in place. However, his goal was not to propose general solutions, but rather (a) to plant the seeds of doubt concerning the existing maps and (b) to reevaluate the territorial extent of some larger groups, such as the Serbs. Sax pronounced Bosnian Moslems a separate nation, in a move with long-term consequences. Because the Moslems were now different from the Serbs, the moral grounds of the Serbs for unification were undermined. At the same time, thanks to Sax's "discovery," things did not look as reprehensible when the multinational Austria-Hungary assumed the protection of another small nation. After the Treaty of Berlin in 1878, Austria occupied Bosnia and Sanjak (and annexed them in 1908). Serbia acquiesced in a secret treaty of 1881. Thereafter, Serbia averted her eyes from the Adriatic and glanced at the Aegean.

Serbian Maps

The maps of the next twenty years merely argued for the already established ideas. However, Sax's techniques did not remain unnoticed; they seem a major inspiration for the ethnographic "revolution" carried through by the prominent Serbian geographer Cvijic. Several features of Cvijic's approach contributed to the success of his maps. First, Cvijic was able to portray himself as an impartial scholar.[24] The claims that he advanced were indirect; they were not to the immediate advantage of any of the sides of the dispute. The categories he introduced appeared balanced. Second, Cvijic was modest and only took gradual steps toward his goal.

Cvijic published a series of four maps between 1906 and 1918. The first one only delineated the ethnographic border between Slavs, Albanians, and Greeks (fig. 4)[25] The position was similar to those of Safarik, Lejean, and other widely accepted notions. Cvijic classified Bulgarians and Serbs in Macedonia as one group: "Serb and Bulgarians," without fixing a frontier between them. Similarly, the Pomaks were described as "Moslem Bulgarians and Serbs."

The next step in the Serb geographer's work was the introduction of the category of "Macedo-Slavs" in a pamphlet of 1906, originally entitled "Remarks on the Ethnography of Macedo-Slavs," and in a map of 1908.[26] These lived in Macedonia in territories previously depicted as Bulgarian. The fact that different maps clashed on the population of this province was not necessarily an expression of the bias of some and the truthfulness of others (namely, the pro-Bulgarian maps). It was, according to Cvijic, an expression of the fact that the Slav population in Macedonia was so hopelessly mixed that this population indeed did not belong to any of the major Slav nations. However, it was not a new nation either. Rather, "The Macedo-Slavs, in their mass, do not have any national self-consciousness—neither Serbian, nor Bulgarian—but only the predisposition to easily adopt such sentiments and become Bulgarians or Serbs."[27] Notice that the idea of "Macedo-Slavs" is not a direct expression of a Serb claim. However, it serves a much subtler goal to convey two notions: (1) It divorces the Slavs from Macedonia—widely perceived as Bulgarian—from the rest of the Bulgarian nation and (2) It opens the possibility for future Serb claims, as the population in question is ethnically neutral.

Another malleable category appearing on this map is ethnically converted populations, for instance, the Albanized Serbs ("Arnauts"). The overt normativity of this category openly invites correction of past (unjust) processes. Likewise, the concept of amorphous Macedo-Slavs openly invites action, an introduction of form to matter. Nowhere does the ethnographic discourse on Macedonia stray further from scientific objectivity than in these two ideas. Under the guise of description, they contain prescriptions for action.

One interesting episode of that period, that illustrates the social skills and scholarly posture of Cvijic, as well the power of his maps, was his audience with the Bulgarian Czar Ferdinand. During their meeting, Cvijic managed to impress the monarch with his scholarship to such an extent that the latter adopted the ideas from Cvijic's map as the state of the art research on Macedonia. The czar consequently agreed to a territorial compromise with Serbia that enabled the Serbo-Bulgarian alliance against Turkey.[28]

The third map, of 1913, confirms the findings of the second one. The ethnographic frontier of the reduced Bulgarians remains the same. The Serb ethnographic frontier is pushed southward, including in its territory the region southwest of Skopje. The "Macedo-Slavs" are "compensated" by the attribution of previously mixed Graeco-Macedo-Slav territories solely to them.

Finally, the map of 1918 completes the maneuver by collapsing the Serbs and the "Macedo-Slavs" into one category—*Serbs and Macedo-Slavs*.[29] Additionally, the frontier between Serbs and Macedo-Slavs and Bulgarians is pushed eastward at the expense of Bulgarians. Much to the astonishment of the Bulgarians, targets of Cvijic's cartographic operations the world's perception was altered so as to regard the population of Macedonia as different from Bulgarian.[30] Furthermore, Cvijic almost succeeded in convincing the geographic community, as well as the international public, that those who were believed to be Bulgarians but who turned out to be Macedo-Slavs are indeed Serbs (or, potential Serbs). As a result, the southern frontier of the Serbian nation was advanced as far as the Gulf of Salonika.

Cvijic's maps were so successful in expanding the Serbian ethnographic frontier that the Serbian annexation of Macedonia appeared legitimate. At the Peace Conference in Paris, the position of Serbia was strong and its territorial claims were accepted as justified. Ironically, the former Bulgarians—the Macedo-Slavs—were not even recognized as a minority, as they were no longer considered a nationality due to Cvijic's maps. Thus, Serbian assimilationist efforts in Macedonia were given a free hand.

III.

In this paper, I have argued that geographic maps cannot secure historical title to territory by virtue of traditional occupation. The maps owe their success in generating moral claims to territory to the prevalence of two theses: the distinguishability thesis that nations are entities, distinguishable according to reliable objective criteria and the representation thesis that the maps are scientific representation of established facts about an objective reality. Upon examination, both theses turned out to be false and the work of human geographers was stripped of the pretense of objectivity and impartiality. Despite the failure of objectivity and impartiality, the work of geographers has an important impact on national identity. And the power of maps to shape national imagination is no weaker if the image diverges from reality.

The mapping of Macedonia by Cvijic constitutes a particularly articulate example of using maps to give direction to a national imagination and, at the same time, to contrive a legal or moral claim on the grounds of traditional occupation. Contrary to the contemporary beliefs that the inhabitants of Macedonia are Bulgarians, Cvijic introduces in a series of maps the idea of "Macedo-Slavs," an amorphous Slav population that is neither Serb nor Bulgarian but may at some time merge with either nation. Thus, he divorces the Slavs from Macedonia from the Bulgarian nation and opens the possibility for future Serb claims. His final map completes the project by depicting Serbs and Macedo-Slavs as one ethnic group. The subtlety of Cvijic's work convinces the public that he is an impartial scholar, an impression that reinforces his project. As a result, the Serbian hold on Macedonia for several decades is accepted as legitimate. It would not be implausible to generalize that Cvijic's project is exemplary of the project of ethnic and national geographers at large.

Notes

1. Yehuda Bloom, *Historic Titles in International Law* (The Hague: Nijhoff, 1965).

2. An early proponent of conquest was Thomas Hobbes in his *Leviathan* (New York: Collier, 1962), chapter 20, 151–155. His account prompted the early criticism of that view by John Locke in his *Two Treatises of Government* (New York: Mentor Book, 1963), chapter 16, §§175–96, 431–44.

3. Among the contemporary political philosophers who accept the traditional occupation view are Harry Beran, "A Liberal Theory of Secession," *Political Studies* 32 (1984): 21–31 and Avishai Margalit and Joseph Raz, "National Self-Determination," *Journal of Philosophy* 87, 9 (September 1990): 439–61.

4. Although this argument is not discussed in the elaborate form I give it, it has been widely accepted. A contemporary example is the position of Margalit and Raz, who accept premises 1 and 2 in their seminal article.

5. The claim expressed in premise 3 is a common place for cartography. Compare the following statement of the British Cartographic Society, "Cartography is the science and technology of analyzing and interpreting geographic relationships, and communicating the results by means of maps" (J.B. Harley, "Deconstructing the Map," in *Human Geography: An Essential Anthology*, ed. J. Agnew, D. Livingstone, and A. Rogers [Oxford: Basil Blackwell, 1996] 424).

6. Often, this position is adopted in a milder form, limiting the entitlement to established, "classic" nations.

7. See, for instance, Karl Marx and Frederick Engels, *Collected Works*, vol. 5, *German Ideology* (New York: International Publishers, 1976), 59.

8. "The 'facts' of geography, presented often as facts of nature, can be used to justify imperialism, neo-colonial domination, and expansionism." David

Harvey, "On the History and Present Condition of Geography: An Historical Materialist Manifesto," *Professional Geographer* 36, 1984, pp. 11–18.

9. J. B. Harley, "Deconstructing the Map," in *Writing Worlds: Discourse, Text and Metaphor in the Representation of Landscape*, ed. T. Barnes and J. Duncan (London: Routledge, 1992), pp. 231–47.

10. I take "objective" to mean something independent of the mind and will of persons. This is the standard philosophical notion, as the following quote from a reference book suggests: "Something is objective if it exists, and is the way it is, independently of any knowledge, perception, conception or consciousness there may be of it" (Jonathan Dancy and Ernest Sosa, eds., *A Companion to Epistemology* (Oxford: Basil Blackwell, 1992), 310. Scientific endeavor, on the other hand, aims at universal and impartial explanations.

11. See, for instance, Hegel's idea that a nation state is a "self-dependent organism" and because it is "a mind objectified, it is only as one of its members that the individual himself has objectivity, genuine individuality, and ethical life." G. W. F. Hegel, *Philosophy of Right*, trans. T. M. Knox (Oxford: Oxford University Press, 1967), 156.

12. Of course, the stronger account implies the weaker. Therefore, if it is refuted, the stronger one is refuted, too.

13. Formally put, family resemblance among nations holds if and only if there is a finite set of objective properties of nations, and each two nations share at least one property belonging to this set.

14. If the argument that family resemblance will stir the practical difference between partial and impartial views seems insufficient, here is another one. Different factors are parts of different nations' definitions because distinctive historical circumstances have shaped the formation of nations. However, what will enter as a part of the definition of a nation is a matter of self-definition, imputed by founding representatives of the nation itself. Therefore, to accept the results of the application of such a self-definition is indeed to accept a design of one of the parties but would not be to accept an objective fact.

15. I will not dwell on this aspect of the geographers' practice. It is discussed prominently by Louis Martin, *Portrait of the King*, trans. Martha Houle (Minneapolis: University of Minnesota Press, 1988), 164–79.

16. Henry R. Wilkinson, *Maps and Politics: A Review of the Ethnographic Cartography of Macedonia* (Liverpool: The University Press, 1951).

17. This is the view expressed in J. P. Fallmeyer's work on Greece and Müller's map of 1842.

18. Pavel Josef Safarik, *Slovansky narodopis*, 4th ed. (Prague: Nakl. Ceskoslovenske akademie, 1955). The map is appended in a pocket. For Boué see Wilkinson, *Maps and Politics*, 44.

19. Wilkinson, *Maps and Politics*, 64.

20. Eduard Stanford, *Ethnographicos chartes tes europaikes Tourkias kai Hellados* (Athens: 1877).

21. This opinion was originally published as a letter in the *London Times*, 30 September 1903; reproduced as "Who the Macedonians Are," in *The Case for an Autonomous Macedonia*, ed. Christ Anastasoff (St. Louis, MO.: Pearlstone, 1945), 36.

22. Wilkinson, *Maps and Politics*, 80.

23. Some examples of nations according to Sax's map are Orthodox Bulgarians, Catholic Bulgarians, Pomaks (Moslem Bulgarians), Bulgarian Uniates, Orthodox Albanians, Catholic Albanians, Moslem Albanians, and so on. See Wilkinson, *Maps and Politics*.

24. In a British publication, Cvijic was introduced as "one of those individuals too rare, especially in the Balkans, who are able to subordinate their patriotism to the cause of scientific exactitude" (Wilkinson, *Maps and Politics*, 148). Cvijic's international reputation as a distinguished geographer was untainted. For instance, in Britain, he was awarded the Patron's Gold Medal of the Royal Geographical Society in part for his study of the human geography of the Balkans (Peter J. Taylor, *Political Geography: World-Economy, Nation-State and Locality*, 2nd ed. [Essex, U.K.: Longman, 1989], 187).

25. Wilkinson, *Maps and Politics*, 147.

26. Jovan Cvijic, *Makedonskie Slaviane: Etnograficheskiia Izsliedovaniia* (Petrograd, 1906). In a later reprint entitled "Macedo-Slavs," Cvijic reminisces that the first publication of the text took place in *Die Zeit* in 1903. It was published also in French (two editions), English (two editions), Russian (two editions), and in the Bulgarian and Czech languages (Cvijic, *Govori i Clanci* [Belgrade, 1921], 193n). For the 1908 map, see Cvijic, *Aneksija Bosne i Hercegovine i Srpski problem*, with two maps (Belgrade, 1908). Interestingly, in this publication in Serbo-Croatian, the ethnographic map is in French (!).

27. Cvijic, *Makedonskie Slaviane*, 3.

28. "The line of partition of which the treaty spoke corresponded fully with the ethnographic conclusions of the learned geographer, Mr. Tsviyits; conclusions which made a profound impression on Czar Ferdinand on the time of his interview with Mr. Tsviyits. It was these conclusions which made the Czar decide to accept the compromise" (*Report of the International Commission of Inquiry into the Balkan Wars* [Washington, D.C.: Carnegie Endowment for International Peace, 1914] 44). Tsviyits here is Cvijic.

29. Wilkinson, *Maps and Politics*, 316.

30. For instance, six years later, Isaiah Bowman accepts Cvijic's thesis on Macedonia in his *The New World: Problems in Political Geography*, 4th ed. (Yonkers, N.Y.: World Book Co., 1924), 360, plate 2.

Index

About the Editors and Contributors

The Editors

Andrew Light is an assistant professor in the Department of Philosophy at The University of Montana, Missoula.

Jonathan M. Smith is an associate professor in the Department of Geography at Texas A. & M. University, College Station.

The Contributors

Edward S. Casey is a professor in the Department of Philosophy at the State University of New York, Stony Brook.

Ian Chaston is director of studies in the Faculty of Business at the University of Plymouth, U.K.

Edward Dimendberg is a philosopher and editor at The University of California Press.

Matthew Gorton is research officer in the Department of Agricultural Economics at Wye College, University of London, U.K.

John Gulick is a Ph.D. candidate in the Department of Sociology at the University of California, Santa Cruz.

Jean Hillier is an associate professor in the Department of Urban and Regional Planning at Curtain University of Technology, Perth, Australia.

Ted Kilian is a graduate student in the Department of Geography at Rutgers University, New Brunswick, New Jersey.

Hugh Mason is on the faculty of the Department of Geography at the University of Portsmouth, U.K.

Mario Pascalev is a Ph.D. candidate in the Department of Philosophy at Bowling Green State University, Ohio.

Neil Smith is a professor in the Department of Geography and acting director of the Center for the Critical Analysis of Contemporary Culture at Rutgers University, New Brunswick, New Jersey.

John Stevenson is a lecturer in the Department of Liberal Education at Columbia College, Chicago, and an instructor in the Department of Philosophy at Roosevelt University, Chicago.

Mary Ann Tétreault is a professor in the Department of Political Science at Iowa State University of Science and Technology, Ames.

Luke Wallin is an associate professor of English at the University of Massachusetts, Dartmouth, a Senior Research Associate at The Center for Policy Analysis, and, presently, Visiting Fulbright Professor at University College, Dublin, Ireland.

John White is a lecturer in the Faculty of Business at the University of Plymouth, U.K.

Style and Submission Guide
Philosophy and Geography

Philosophy and Geography is a peer reviewed annual with each volume focusing on a specific theme. Each issue addresses a topic of mutual interest to philosophers and geographers. The annual is edited by Andrew Light and Jonathan M. Smith, in consultation with the editorial board, and published by Rowman & Littlefield Publishers, Inc. All material submitted to the editors is subjected to peer review by members of the editorial board, Associate Editors of the journal, or others, serving at the behest of the editors. (Themes and deadlines for upcoming issues are listed at the front of this volume.)

Length

Authors should aim for manuscripts of about 10,000 words, including the notes. If you are using a type size and font similar to the one in this letter, this will yield a double spaced manuscript of about thirty pages. Shorter and longer manuscripts will be considered, but only extraordinary circumstances will justify acceptance of manuscripts of less than 6,000 or more than 12,000 words. As this length includes the notes, authors are urged to limit notes to citation of works directly relevant to your argument.

Submission

Authors should send three copies of their manuscript to:

> Andrew Light, Co-Editor
> *Philosophy and Geography*
> Department of Philosophy
> The University of Montana
> Missoula, MT 59812

Each copy must be single-sided, double-spaced, in a large type size, with wide margins (one inch margins preferred). Illustrations submitted with the final draft of accepted manuscripts must be camera ready. Please do not send bound manuscripts. Once an article has been accepted, authors must submit a disk copy of their manuscript.

Notation Style

Authors should follow the *Chicago Manual of Style* and use American spelling. Notes will be printed as endnotes, and should be used judiciously. Do not use more than one note in a single sentence, and whenever possible group all of the citations and asides from an entire paragraph in a single note. Here are examples of some common citations:

Book: Clarence J. Glacken, *Traces on the Rhodian Shore* (Berkeley and Los Angeles: University of California Press, 1967), xiii.

Volume Chapter: Roger J. H. King, "Relativism and Moral Critique," in *The American Constitutional Experiment*, ed. David M. Speak and Creighton Peden (Lewiston, N. Y.: The Edwin Mellen Press, 1991), 145–64.

Journal: Eric Katz, "The Call of the Wild: The Struggle Against Domination and the Technological Fix," *Environmental Ethics*, 14, no. 3 (1992): 271.

Newspaper: Timothy Egan, "Unlikely Alliances Attack Property Rights Measures," *New York Times*, 15 May, 1995, A1.

The full citation should be given only in the first note in which a work is cited. A short form should be used in all subsequent notes. Short forms of the works cited above might appear as follows:

Book: Glacken, *Traces on the Rhodian Shore*, 372.

Volume Chapter: King, "Relativism and Moral Critique," 146.

Journal: Katz, "The Call of the Wild," 272

Newspaper: Egan, "Unlikely Alliances."

Originality

Authors are asked to submit a letter with their manuscript stating that the material in the manuscript has not been published elsewhere, that it is not presently under consideration by another publication, and that it will not be submitted for consideration by another publication until the author has been notified of the final decision of the editors of *Philosophy and Geography*. Upon publication, Rowman & Littlefield will possess the copyright.

Editorial Policy

Philosophy and Geography is on an accelerated production schedule. Papers accepted in February appear in print in the Fall of the same year (paperback in October, hard cover in November), and the editorial work occurs in a much shorter period. It may not be possible to return either the final copy-edited manuscript or the page proofs to the authors for approval. The editors will contact an author if there appears to be a need for an extensive, significant, or objectionable change to his or her manuscript, but they will not seek an author's consent to make minor alterations, additions, or deletions. These decisions are made at the editor's discretion.